Signals and Communication Technology

For further volumes:
http://www.springer.com/series/4748

Feng Li

Interference Cancellation Using Space-Time Processing and Precoding Design

 Springer

Feng Li
Department of Operations Research
  and Information Engineering
Cornell University
Ithaca 14853 NY
USA

ISSN 1860-4862
ISBN 978-3-642-42848-7   ISBN 978-3-642-30712-6 (eBook)
DOI 10.1007/978-3-642-30712-6
Springer Heidelberg New York Dordrecht London

To my family

Preface

In this book, we study the interference cancellation and detection problem in multiantenna multi-user scenario using precoders. The goal is to utilize multiple antennas to cancel the interference without sacrificing the diversity or the complexity of the system.

First, we consider the case with two users and one receiver when users know each other channels. Before, in the literature, it was shown how a receiver with two receive antennas can completely cancel the interference of two users and provide a diversity of two for users with two transmit antennas.We propose a system to achieve the maximum possible diversity of four with low complexity. Our main idea is to design precoders, using the channel information, to make it possible for different users to transmit over orthogonal spaces. Then, using the orthogonality of the transmitted signals, the receiver can separate them and decode the signals independently. We also extend our scheme to any number of antennas and analytically prove that the system provides full diversity to both users.

However, the above scheme only works for two users. So we extend the scheme to more than two users. In other words, we propose a system to achieve interference cancellation and full diversity with low complexity for any number of users. Then, we extend the results to any number of users with any number of transmit and receive antennas. Our main idea is to design precoders, using the channel information, to make it possible for different users to transmit over orthogonal directions. Then, using the orthogonality of the transmitted signals, the receiver can separate them and decode the signals independently. We also analytically prove that our system provides full diversity to each user.

In practice, perfect channel information is not available, so we design precoders for two users with two transmit antennas and one receiver with two receive antennas using quantized feedback. We propose to construct codebook using Grassmannian line packing. By choosing precoders from the codebook properly, our proposed scheme can cancel the interference for each user. Also we analytically prove that our system can achieve full diversity for each user. Then we extend our scheme to any number of transmit and receive antennas. Simulation results confirm our analytical proof and show that our scheme can serve as a bridge between a system with no feedback and a system with perfect feedback.

Finally, we investigate how to send codewords without interference with full diversity and low decoding complexity for X channels. We assume that we have two transmitters and two receivers. Each transmitter sends different codewords to each receiver at the same time. We propose our precoding and decoding schemes such that each receiver can get the desired signals from each transmitter without any interference. We show that our proposed scheme can provide full diversity for transmitted signals. Also our decoding complexity is low. To our best knowledge, this is the first scheme which can achieve interference-free transmission and full diversity for any transmitted codeword in X channel when all the users transmit at the same time.We also show that under certain conditions, our proposed scheme can be extended to a general case with any number of transmitters and receivers each with any number of antennas.

Ithaca, New York, February 2012   Feng Li

Acknowledgments

The author would like to thank Professor Hamid Jafarkhani from the University of California, Irvine for his strong support and help in writing this book.

Contents

Chapter 1
Introduction

1.1 Interference Cancellation and Detection for Multiple Access Channel with Perfect Feedback

Multi-user detection schemes with simple receiver structures have received a lot of attention lately. Multiple transmit and receive antennas have been used to increase rate and improve the reliability of wireless systems. In this chapter, we consider a multiple-antenna multi-access scenario where receive antennas are utilized to cancel the interference. When there is channel information at the transmitter, in [1], multiple antennas have been used to suppress the interference from other users. They show that one can decode each user separately by using enough number of receive antennas. More specifically, for J users equipped with N transmit antennas, they show how to cancel the interference using NJ receive antennas. To reduce the number of required receive antennas, [2] provides an interference cancellation method for users with 2 transmit antennas. The method is based on the properties of orthogonal space-time block codes (OSTBCs) [3] and requires less number of receive antennas, i.e. as many as the number of users. The work was extended to a higher number of transmit antennas but only for $J = 2$ users in [4]. The common theme of the work in [2, 4] is the utilization of the properties of the orthogonal designs [3] at the transmitter to cancel the interference at the receiver. Unfortunately, the method does not work for a general case of complex constellations, $N > 2$ transmit antennas, and $J > 2$ users [5]. In fact, [5] proves that such an extension using orthogonal designs is impossible. Instead, [5] suggests a method based on quasi-orthogonal space-time block codes (QOSTBCs) [6]. The main complexity tradeoff between OSTBCs and QOSTBCs is the symbol-by-symbol decoding versus pairwise decoding. Therefore, by a moderate increase of decoding complexity, [5] extends the above multi-user detection schemes to any constellation, any number of users, and any number of transmit antennas. Performance analysis of these systems in terms of signal-to-noise ratio (SNR) is available in [7, 8]. Further, it is shown in [9] that for $M \geq J$ receive antennas, the diversity of each user is equal to NM using maximum-likelihood detection and $N(M - J + 1)$ using low-complexity array-processing schemes. Note that the

F. Li, *Interference Cancellation Using Space-Time Processing and Precoding Design*,
Signals and Communication Technology, DOI: 10.1007/978-3-642-30712-6_1,
© Springer-Verlag Berlin Heidelberg 2013

complexity of the maximum-likelihood detection increases exponentially as a function of the number of antennas, the number of users, and the bandwidth efficiency (measured in bits per channel use). Therefore, usually it is not practical.

Applying a linear transformation to the codeword before transmission is called linear precoding. When there are some feedback available at the transmitter, precoding techniques can be used to improve the system performance. For example, in a point-to-point multiple-input multiple-output (MIMO) channel, the performance of the system can be improved significantly using precoders [10–16]. In a multi-antenna multiple access channel, one can utilize the channel information to improve the performance of the system while achieving interference cancellation using array processing methods mentioned above. For example, in [8], post-processing SNR is maximized for a given linear receiver by selecting the QOSTBC with the minimum quaternionic angle as well as realizing interference cancellation. In [17], limited feedback is used to adapt the phase of a transmitted signal to improve the performance of the system.

The common goal and the main characteristics of the above multi-user systems are the small number of required receive antennas and the low complexity of the array-processing decoding. A receiver does not need more than J receive antennas and the decoding is symbol-by-symbol or pairwise using low-complexity array-processing methods. One drawback, however, is that if we demand low complexity, the maximum diversity of NM is not achievable. Our motivation is to utilize the channel information at the user transmitters to increase the diversity of the system while keeping the low complexity of the decoding. In other words, unlike the above-mentioned methods, we do not use receive antennas to cancel the interference. Instead, we use the channel information at the transmitter to design precoders that align different groups of signals along orthogonal directions. As a result, interference suppression is achieved without utilizing the receive antenna resources and therefore full diversity is achieved naturally.

1.2 Interference Cancellation and Detection for Multiple Access Channel with Quantized Feedback

Although the performance of the scheme in [18] is better than that of the former proposed schemes, perfect channel information is needed at transmitters. It is not practical in reality. We investigate the use of limited feedback to achieve interference cancellation as well as full diversity. Limited feedback has been used extensively in the case of the single-user MIMO systems. It has been shown that the capacity and performance of the point-to-point MIMO systems can be increased significantly using limited feedback [10–16]. There are few examples of multi-antenna multi-user systems with limited feedback in the literature. In [8], post-processing SNR is maximized for a given linear receiver by selecting the QOSTBC with the minimum quaternionic angle as well as realizing interference cancellation. In [17], limited feedback is utilized to adapt the phase of a transmitted signal and improve the performance of the system. However, to the best of our knowledge, there is no result showing how to achieve full diversity and interference cancellation for each user using limited feedback. A naive way is to quantize the result in [18] directly. But this will not work because the scheme in [18] relies on the perfect channel information and thus perfect orthogonality between the signal vectors of the two users. Simply quantizing the results will destroy the perfect orthogonality and thus cannot achieve full diversity. In this book we investigate how to use quantized feedback to achieve full diversity as well as interference cancellation. Our results show that even with quantized feedback, full diversity and interference cancellation are possible by using our proposed scheme. Also our decoding complexity is the lowest to our best knowledge. By increasing the number of feedback bits, the performance of our proposed scheme will approach the performance of the scheme with perfect feedback in [18]. So our proposed scheme can serve as a bridge between the schemes with no feedback and perfect feedback.

1.3 Interference Cancellation and Detection for X Channels

When there are two users each transmitting different codewords to two receivers simultaneously, a scenario known as X channels, multiple antennas can be utilized to increase the date rate as discussed in the existing literature. For example, the schemes proposed in [19, 20] can achieve the highest multiplexing gain with no or partial cooperation between users. References [21, 22] provide the achievability as well as converse results for the degrees of freedom region of a MIMO X channel using a technique called interference alignment when perfect channel knowledge is available to all transmitters and receivers. Capacity region and Relay-Aided X channels are discussed in [23, 24].

The main emphasis of the above papers has been the maximum multiplexing gain. But, in most cases, these schemes achieve a diversity of one. On the other hand, in a system with limited complexity and delay constraints, reliability, in terms of error probability or diversity, is also important. Interference cancellation methods have been proposed to suppress the interference from other users and reduce the error probability in multiple access channels with limited delay and complexity [2, 4, 5, 9, 25]. Unfortunately, these interference cancellation methods cannot be used in X channels.

We investigate the following two problems: (1) how to realize interference-free transmission for each user to reduce the error probability and the decoding complexity. (2) how to achieve the highest possible diversity to improve the transmission quality in X channels. To the best of our knowledge, the only way to achieve full diversity for each user in X channels is to jointly decode the codewords from all users using maximum-likelihood decoding. The decoding complexity of such a scheme is very high. Also, each receiver will have access to the information of the other receiver which may not be desirable. Our proposed precoding and decoding scheme can successfully cancel the interference without sacrificing diversity. Thus we can decode codewords for each user separately and the decoding complexity is reduced largely. To our knowledge, this is the first scheme that can achieve interference cancellation and full diversity for X channels, considering limited delay and complexity for practical constellations.

References

1. Tarokh, V., Naguib, A., Seshadri, N., Calderbank, A.R.: Combined array processing and space-time coding. IEEE Trans. Inform. Theory **45**, 1121–1128 (1999)
2. Naguib, A.F., Seshadri, N., Calderbank, A.R.: Applications of space-time block codes and interference suppression for high capacity and high data rate wireless systems. In: Proceedings of 32nd Asilomar Conference on Signals, Systems and Computers, pp. 1803–1810 (1998).
3. Tarokh, V., Jafarkhani, H., Calderbank, A.R.: Space-time block codes from orthogonal designs. IEEE Trans. Inform. Theory **45**, 1456–1467 (1999)
4. Al-Dhahir, N., Calderbank, A.R.: Further results on interference cancellation and space-time block codes. In: Proceedings of 35th Asilomar Conference on Signals, Systems and Computers, pp. 257–262 (2001).
5. Kazemitabar, J., Jafarkhani, H.: Multiuser interference cancellation and detection for users with more than two transmit antennas. IEEE Trans. Commun. **56**(4), 574–583 (2008)
6. Jafarkhani, H.: A quasi-orthogonal space-time block code. IEEE Trans. Commun. **49**(1), 1–4 (2001)
7. Sirianunpiboon, S., Howard, S.D., Calderbank, A.R.: Diversity gains across line of sight and rich scattering environments from space-polarization-time codes. In: IEEE Information Theory Workshop on Information Theory for, Wireless Networks, pp. 1–5 (2007).
8. Tan, C.W., Calderbank, A.R.: Multiuser detection of Alamouti signals. IEEE Trans. Commun. **57**(7), 2080–2089 (2009)
9. Kazemitabar, J., Jafarkhani, H.: Performance analysis of multiple-antenna multi-user detection. In: Proceedings of 2009 Workshop on Information Theory and its Applications (2009).
10. Scaglione, A., Stoica, P., Barbarossa, S., Giannakis, G., Sampath, H.: Optimal designs for space-time linear precoders and decoders. IEEE Trans. Sig. Process. **50**(5), 1051–1064 (2002)
11. Love, D., Heath, R.J.: Limited feedback unitary precoding for orthogonal space-time block codes. IEEE Trans. Sig. Process. **53**(1), 64–73 (2005)
12. Ghaderipoor, A., Tellambura, C.: Optimal precoder for rate less than one space-time block codes. Proceedings of IEEE International Conference on Communications, Glasgow, Scotland, In (2007)
13. Sampath, H., Paulraj, A.: Linear precoding for space-time coded systems with known fading correlations. IEEE Commun. Lett. **6**(6), 239–241 (2002)
14. Jongren, G., Skoglund, M., Ottersten, B.: Combining beamforming and orthogonal space-time block coding. IEEE Trans. Inf. Theory **48**, 611–627 (2002)
15. Liu, L., Jafarkhani, H.: Application of quasi-orthogonal space-time block codes in beamforming. IEEE Trans. Sig. Process. **53**(1), 54–63 (2005)
16. Ekbatani, S., Jafarkhani, H.: Combining beamforming and space-time coding using quantized feedback. IEEE Trans. Wireless Commun. **7**(3), 898–908 (2008)
17. Wu, Y.Y., Calderbank, A.R.: Code diversity in multiple antenna wireless communication. In: Proceedings of IEEE International Symposium on Information Theory, pp. 1078–1082, Toronto, Canada (2008).
18. Li, F., Jafarkhani, H.: Interference cancellation and detection using precoders. Proceedings of IEEE International Conference on Communications, Dresden, Germany, In (2009)

19. Maddah-Ali, M.A., Motahari, A.S., Khandani, A.K.: Communication over MIMO X channels: interference alignment, decomposition, and performance analysis. IEEE Trans. Inf. Theory **54**(8), 3457 3470 (2008)
20. Devroye, N., Sharif, M.: The multiplexing gain of MIMO X-channels with partial transmit side-information. In: Proceedings of the IEEE International Symposium on Information Theory (ISIT '07), Nice, France (2007).
21. Cadambe, V.R., Jafar, S.A.: Interference Alignment and the Degrees of Freedom of Wireless X Networks. IEEE Trans. Inf. Theory **55**(9), 3893–3908 (2009)
22. Jafar, S.A., Shamai, S.: Degrees of Freedom Region for the MIMO X Channel. IEEE Trans. Inf. Theory **54**(1), 151–170 (2008)
23. Koyluoglu, O.O., Shahmohammadi, M., El Gamal, H.: A new achievable rate region for the discrete memoryless X channel. In: Proceedings of 2009 IEEE International Symposium on Information Theory (ISIT 2009), pp. 2427–2431 (2009).
24. Nourani, B., Motahari, A.S., Khandani, A.K.: Relay-aided interference alignment for the quasi-static X channel. In: Proceedings of IEEE International Symposium on Information Theory 2009, pp. 1764–1768 (2009).
25. Li, F., Jafarkhani, H.: Multiple-antenna interference cancellation and detection for two users using precoders. IEEE J. Sel. Top. Sig. Process. **3**(6), 1066–1078 (2009)

Chapter 2
Interference Cancellation and Detection for MAC with Two Users

2.1 Channel Model

In this chapter, we assume a quasi-static flat Rayleigh fading channel model for the channel as shown in Fig. 2.1. The path gains are independent complex Gaussian random variables and fixed during the transmission of one block. There are two users each with two transmit antennas and one receiver with two receive antennas.

At the first two time slots, the channel matrices for Users 1 and 2 are

$$\mathbf{H} = \begin{pmatrix} h_{11} & h_{12} \\ h_{21} & h_{22} \end{pmatrix}, \quad \mathbf{G} = \begin{pmatrix} g_{11} & g_{12} \\ g_{21} & g_{22} \end{pmatrix} \tag{2.1}$$

respectively, where h_{ij} and g_{ij} are i.i.d. $CN(0, 1)$.

At the first two time slots, Users 1 and 2 transmit Alamouti codes

$$\mathbf{C} = \begin{pmatrix} c_1 & -c_2^* \\ c_2 & c_1^* \end{pmatrix}, \quad \mathbf{S} = \begin{pmatrix} s_1 & -s_2^* \\ s_2 & s_1^* \end{pmatrix} \tag{2.2}$$

respectively. At time slots 1 and 2, the received signals are respectively denoted by

$$\mathbf{y}^1 = \begin{pmatrix} y_1^1 \\ y_2^1 \end{pmatrix}, \quad \mathbf{y}^2 = \begin{pmatrix} y_1^2 \\ y_2^2 \end{pmatrix} \tag{2.3}$$

We assume that the transmitter and receiver know the channel information perfectly. Let

$$\mathbf{A}^1 = \begin{pmatrix} a_{11}^1 & a_{12}^1 \\ a_{21}^1 & a_{22}^1 \end{pmatrix}, \quad \mathbf{A}^2 = \begin{pmatrix} a_{11}^2 & a_{12}^2 \\ a_{21}^2 & a_{22}^2 \end{pmatrix} \tag{2.4}$$

denote the precoders of User 1 at time slots 1 and 2, respectively. Also,

F. Li, *Interference Cancellation Using Space-Time Processing and Precoding Design*, Signals and Communication Technology, DOI: 10.1007/978-3-642-30712-6_2, © Springer-Verlag Berlin Heidelberg 2013

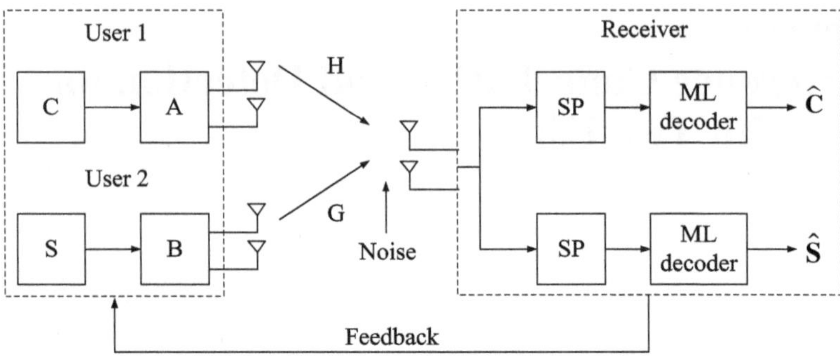

Fig. 2.1 Channel model

$$\mathbf{B}^1 = \begin{pmatrix} b_{11}^1 & b_{12}^1 \\ b_{21}^1 & b_{22}^1 \end{pmatrix}, \quad \mathbf{B}^2 = \begin{pmatrix} b_{11}^2 & b_{12}^2 \\ b_{21}^2 & b_{22}^2 \end{pmatrix} \tag{2.5}$$

denote the precoders of User 2 at time slots 1 and 2, respectively. We assume $||\mathbf{A}^i||_F^2 = ||\mathbf{B}^i||_F^2 = 1$ in order to satisfy the normalization conditions [1].

Our goal is to design low-complexity precoders to realize interference cancellation and full diversity for each user. The main idea is to design precoders such that the two users transmit over two orthogonal spaces. As a result, the decoders can project the received signals to each of the orthogonal spaces and decode the information of each user without any interference from the other user. Later, we prove that the resulted diversity is full for each user.

2.2 Precoding Design

We first present the precoder design for time slot 1. Then, a similar design strategy for time slot 2 is briefly discussed. We present our precoder design method through the following four steps, which are also illustrated in the flow chart in Fig. 2.2:

Step 1: Deriving the equivalent channel equations:

At time slot 1, the signal model can be written as

$$\mathbf{y}^1 = \sqrt{E_s}\mathbf{H}\mathbf{A}^1 \begin{pmatrix} c_1 \\ c_2 \end{pmatrix} + \sqrt{E_s}\mathbf{G}\mathbf{B}^1 \begin{pmatrix} s_1 \\ s_2 \end{pmatrix} + \mathbf{W}^1 \tag{2.6}$$

At time slot 2, we have

$$\mathbf{y}^2 = \sqrt{E_s}\mathbf{H}\mathbf{A}^2 \begin{pmatrix} -c_2^* \\ c_1^* \end{pmatrix} + \sqrt{E_s}\mathbf{G}\mathbf{B}^2 \begin{pmatrix} -s_2^* \\ s_1^* \end{pmatrix} + \mathbf{W}^2 \tag{2.7}$$

Fig. 2.2 Flow chart for our design

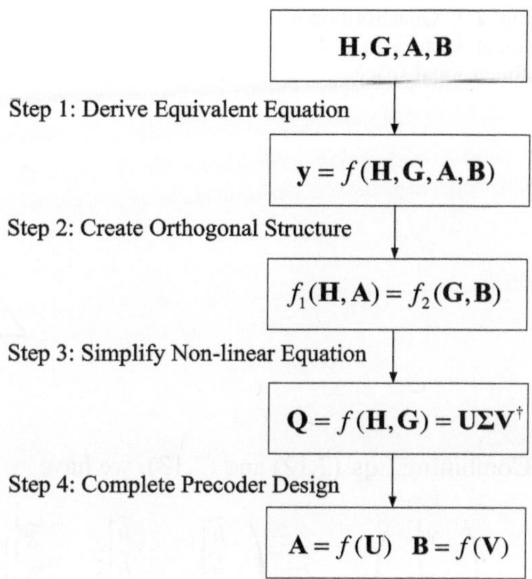

where E_s denotes the total transmit energy of each user and $\mathbf{W}^1 = \begin{pmatrix} n_1^1 \\ n_2^1 \end{pmatrix}$, $\mathbf{W}^2 = \begin{pmatrix} n_1^2 \\ n_2^2 \end{pmatrix}$ denote the noise at the receiver at time slots 1 and 2, respectively. We assume that $n_1^1, n_2^1, n_1^2, n_2^2$ are i.i.d complex Gaussian noises with mean 0 and variance 1. If we let

$$\widehat{\mathbf{H}}^1 = \begin{pmatrix} \widehat{h}_{11}^1 & \widehat{h}_{12}^1 \\ \widehat{h}_{21}^1 & \widehat{h}_{22}^1 \end{pmatrix} = \mathbf{HA}^1 = \begin{pmatrix} h_{11}a_{11}^1 + h_{12}a_{21}^1 & h_{11}a_{12}^1 + h_{12}a_{22}^1 \\ h_{21}a_{11}^1 + h_{22}a_{21}^1 & h_{21}a_{12}^1 + h_{22}a_{22}^1 \end{pmatrix} \quad (2.8)$$

$$\widehat{\mathbf{G}}^1 = \begin{pmatrix} \widehat{g}_{11}^1 & \widehat{g}_{12}^1 \\ \widehat{g}_{21}^1 & \widehat{g}_{22}^1 \end{pmatrix} = \mathbf{GB}^1 = \begin{pmatrix} g_{11}b_{11}^1 + g_{12}b_{21}^1 & g_{11}b_{12}^1 + g_{12}b_{22}^1 \\ g_{21}b_{11}^1 + g_{22}b_{21}^1 & g_{21}b_{12}^1 + g_{22}b_{22}^1 \end{pmatrix} \quad (2.9)$$

$$\widehat{\mathbf{H}}^2 = \begin{pmatrix} \widehat{h}_{11}^2 & \widehat{h}_{12}^2 \\ \widehat{h}_{21}^2 & \widehat{h}_{22}^2 \end{pmatrix} = \mathbf{HA}^2 = \begin{pmatrix} h_{11}a_{11}^2 + h_{12}a_{21}^2 & h_{11}a_{12}^2 + h_{12}a_{22}^2 \\ h_{21}a_{11}^2 + h_{22}a_{21}^2 & h_{21}a_{12}^2 + h_{22}a_{22}^2 \end{pmatrix} \quad (2.10)$$

$$\widehat{\mathbf{G}}^2 = \begin{pmatrix} \widehat{g}_{11}^2 & \widehat{g}_{12}^2 \\ \widehat{g}_{21}^2 & \widehat{g}_{22}^2 \end{pmatrix} = \mathbf{GB}^2 = \begin{pmatrix} g_{11}b_{11}^2 + g_{12}b_{21}^2 & g_{11}b_{12}^2 + g_{12}b_{22}^2 \\ g_{21}b_{11}^2 + g_{22}b_{21}^2 & g_{21}b_{12}^2 + g_{22}b_{22}^2 \end{pmatrix} \quad (2.11)$$

then channel equations (4.8) and (4.9) can be written as

$$\begin{pmatrix} y_1^1 \\ y_2^1 \end{pmatrix} = \sqrt{E_s} \begin{pmatrix} \widehat{h}_{11}^1 & \widehat{h}_{12}^1 \\ \widehat{h}_{21}^1 & \widehat{h}_{22}^1 \end{pmatrix} \begin{pmatrix} c_1 \\ c_2 \end{pmatrix} + \sqrt{E_s} \begin{pmatrix} \widehat{g}_{11}^1 & \widehat{g}_{12}^1 \\ \widehat{g}_{21}^1 & \widehat{g}_{22}^1 \end{pmatrix} \begin{pmatrix} s_1 \\ s_2 \end{pmatrix} + \begin{pmatrix} n_1^1 \\ n_2^1 \end{pmatrix} \quad (2.12)$$

$$\begin{pmatrix} y_1^2 \\ y_2^2 \end{pmatrix} = \sqrt{E_s} \begin{pmatrix} \widehat{h}_{11}^2 & \widehat{h}_{12}^2 \\ \widehat{h}_{21}^2 & \widehat{h}_{22}^2 \end{pmatrix} \begin{pmatrix} -c_2^* \\ c_1^* \end{pmatrix} + \sqrt{E_s} \begin{pmatrix} \widehat{g}_{11}^2 & \widehat{g}_{12}^2 \\ \widehat{g}_{21}^2 & \widehat{g}_{22}^2 \end{pmatrix} \begin{pmatrix} -s_2^* \\ s_1^* \end{pmatrix} + \begin{pmatrix} n_1^2 \\ n_2^2 \end{pmatrix} \quad (2.13)$$

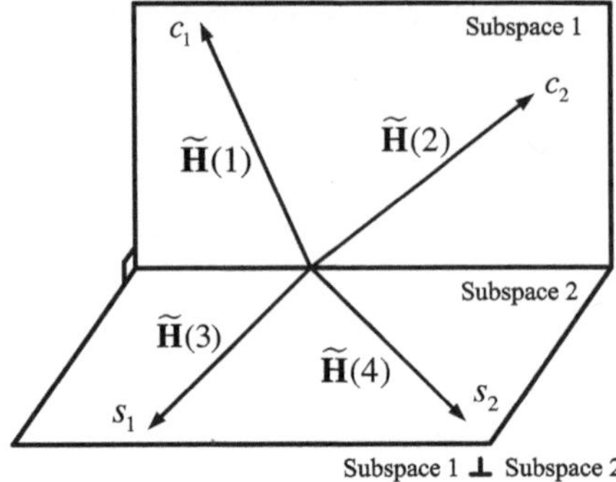

Fig. 2.3 Quasi-orthogonal signal vectors in a 4-dimensional space

Combining Eqs. (2.12) and (2.13), we have

$$
\begin{pmatrix} y_1^1 \\ y_2^1 \\ (y_1^2)^* \\ (y_2^2)^* \end{pmatrix} = \sqrt{E_s} \begin{pmatrix} \widehat{h}_{11}^1 & \widehat{h}_{12}^1 & \widehat{g}_{11}^1 & \widehat{g}_{12}^1 \\ \widehat{h}_{21}^1 & \widehat{h}_{22}^1 & \widehat{g}_{21}^1 & \widehat{g}_{22}^1 \\ (\widehat{h}_{12}^2)^* & -(\widehat{h}_{11}^2)^* & (\widehat{g}_{12}^2)^* & -(\widehat{g}_{11}^2)^* \\ (\widehat{h}_{22}^2)^* & -(\widehat{h}_{21}^2)^* & (\widehat{g}_{22}^2)^* & -(\widehat{g}_{21}^2)^* \end{pmatrix} \begin{pmatrix} c_1 \\ c_2 \\ s_1 \\ s_2 \end{pmatrix} + \begin{pmatrix} n_1^1 \\ n_2^1 \\ (n_1^2)^* \\ (n_2^2)^* \end{pmatrix}
$$

$$(2.14)$$

Equation (2.14) is the equivalent channel equation and we define

$$
\widetilde{\mathbf{H}} = \begin{pmatrix} \widehat{h}_{11}^1 & \widehat{h}_{12}^1 & \widehat{g}_{11}^1 & \widehat{g}_{12}^1 \\ \widehat{h}_{21}^1 & \widehat{h}_{22}^1 & \widehat{g}_{21}^1 & \widehat{g}_{22}^1 \\ (\widehat{h}_{12}^2)^* & -(\widehat{h}_{11}^2)^* & (\widehat{g}_{12}^2)^* & -(\widehat{g}_{11}^2)^* \\ (\widehat{h}_{22}^2)^* & -(\widehat{h}_{21}^2)^* & (\widehat{g}_{22}^2)^* & -(\widehat{g}_{21}^2)^* \end{pmatrix}, \quad \widetilde{\mathbf{n}} = \begin{pmatrix} n_1^1 \\ n_2^1 \\ (n_1^2)^* \\ (n_2^2)^* \end{pmatrix}
$$

$$(2.15)$$

Step 2: Creating the orthogonal structure of signal vectors:

We aim to align signals along several orthogonal vectors to separate them completely. From Eq. (2.14), we know that we have 4 useful symbols of the 2 users. If we can transmit them along 4 orthogonal vectors, it is obvious that we can separate them easily at the receiver. But we know that the 4 dimensional complex orthogonal design does not exist. So we can utilize the quasi-orthogonal design. In other words, we can make the subspace 1 created by the first two columns of matrix $\widetilde{\mathbf{H}}$ orthogonal to the subspace 2 created by the second two columns of matrix $\widetilde{\mathbf{H}}$, as shown in Fig. 2.3. Then at the receiver, we can separate the signals of User 1 from the signals of User 2 easily. This is the main idea of our interference cancellation scheme.

In order to create the quasi-orthogonal structure, first, we let

$$
\mathbf{A}^1(1) = \mathbf{A}^1(2), \quad \mathbf{A}^2(1) = \mathbf{A}^2(2)
$$
$$
\mathbf{B}^1(1) = \mathbf{B}^1(2), \quad \mathbf{B}^2(1) = \mathbf{B}^2(2)
$$

$$(2.16)$$

i.e.,

$$\begin{pmatrix} a_{11}^1 \\ a_{21}^1 \end{pmatrix} = \begin{pmatrix} a_{12}^1 \\ a_{22}^1 \end{pmatrix}, \quad \begin{pmatrix} a_{11}^2 \\ a_{21}^2 \end{pmatrix} = \begin{pmatrix} a_{12}^2 \\ a_{22}^2 \end{pmatrix} \tag{2.17}$$

$$\begin{pmatrix} b_{11}^1 \\ b_{21}^1 \end{pmatrix} = \begin{pmatrix} b_{12}^1 \\ b_{22}^1 \end{pmatrix}, \quad \begin{pmatrix} b_{11}^2 \\ b_{21}^2 \end{pmatrix} = \begin{pmatrix} b_{12}^2 \\ b_{22}^2 \end{pmatrix} \tag{2.18}$$

From Eqs. (2.8), (4.11), (4.15), (2.18), we can easily derive

$$\begin{pmatrix} \widehat{h}_{11}^1 \\ \widehat{h}_{21}^1 \end{pmatrix} = \begin{pmatrix} \widehat{h}_{12}^1 \\ \widehat{h}_{22}^1 \end{pmatrix}, \quad \begin{pmatrix} (\widehat{h}_{12}^2)^* \\ (\widehat{h}_{22}^2)^* \end{pmatrix} = \begin{pmatrix} (\widehat{h}_{11}^2)^* \\ (\widehat{h}_{21}^2)^* \end{pmatrix}$$

$$\begin{pmatrix} \widehat{g}_{11}^1 \\ \widehat{g}_{21}^1 \end{pmatrix} = \begin{pmatrix} \widehat{g}_{12}^1 \\ \widehat{g}_{22}^1 \end{pmatrix}, \quad \begin{pmatrix} (\widehat{g}_{12}^2)^* \\ (\widehat{g}_{22}^2)^* \end{pmatrix} = \begin{pmatrix} (\widehat{g}_{11}^2)^* \\ (\widehat{g}_{21}^2)^* \end{pmatrix} \tag{2.19}$$

For simplicity, (2.14) can be written as

$$\begin{pmatrix} y_1^1 \\ y_2^1 \\ (y_1^2)^* \\ (y_2^2)^* \end{pmatrix} = \sqrt{E_s} \begin{pmatrix} \widehat{h}_{11}^1 & \widehat{h}_{11}^1 & \widehat{g}_{11}^1 & \widehat{g}_{11}^1 \\ \widehat{h}_{21}^1 & \widehat{h}_{21}^1 & \widehat{g}_{21}^1 & \widehat{g}_{21}^1 \\ (\widehat{h}_{12}^2)^* & -(\widehat{h}_{12}^2)^* & (\widehat{g}_{12}^2)^* & -(\widehat{g}_{12}^2)^* \\ (\widehat{h}_{22}^2)^* & -(\widehat{h}_{22}^2)^* & (\widehat{g}_{22}^2)^* & -(\widehat{g}_{22}^2)^* \end{pmatrix} \begin{pmatrix} c_1 \\ c_2 \\ s_1 \\ s_2 \end{pmatrix} + \widetilde{\mathbf{n}} \tag{2.20}$$

Now, we let

$$\begin{pmatrix} \widehat{g}_{11}^1 \\ \widehat{g}_{21}^1 \end{pmatrix} = \eta_1 \begin{pmatrix} -(\widehat{h}_{21}^1)^* \\ (\widehat{h}_{11}^1)^* \end{pmatrix}, \quad \begin{pmatrix} (\widehat{g}_{12}^2)^* \\ (\widehat{g}_{22}^2)^* \end{pmatrix} = \eta_2 \begin{pmatrix} -\widehat{h}_{22}^2 \\ \widehat{h}_{12}^2 \end{pmatrix} \tag{2.21}$$

where η_1 and η_2 are parameters we will determine later. Therefore, (2.20) can be written as

$$\begin{pmatrix} y_1^1 \\ y_2^1 \\ (y_1^2)^* \\ (y_2^2)^* \end{pmatrix} = \sqrt{E_s} \begin{pmatrix} \widehat{h}_{11}^1 & \widehat{h}_{11}^1 & -\eta_1(\widehat{h}_{21}^1)^* & -\eta_1(\widehat{h}_{21}^1)^* \\ \widehat{h}_{21}^1 & \widehat{h}_{21}^1 & \eta_1(\widehat{h}_{11}^1)^* & \eta_1(\widehat{h}_{11}^1)^* \\ (\widehat{h}_{12}^2)^* & -(\widehat{h}_{12}^2)^* & -\eta_2\widehat{h}_{22}^2 & \eta_2\widehat{h}_{22}^2 \\ (\widehat{h}_{22}^2)^* & -(\widehat{h}_{22}^2)^* & \eta_2\widehat{h}_{12}^2 & -\eta_2\widehat{h}_{12}^2 \end{pmatrix} \begin{pmatrix} c_1 \\ c_2 \\ s_1 \\ s_2 \end{pmatrix} + \widetilde{\mathbf{n}} \tag{2.22}$$

Note that, four symbols are transmitted along four columns of matrix $\widetilde{\mathbf{H}}$. The first two columns are orthogonal to the second two columns. So c_1, c_2 and s_1, s_2 are transmitted in two orthogonal subspaces as shown in Fig. 2.3. In this way, we can separate them and achieve interference cancellation for each user at the receiver.

Step 3: Designing low-complexity algorithms to calculate the parameters in the precoders:

In order to get the quasi-orthogonal structure given in (2.22), Eq. (2.21) shows that we need to solve the following equations

$$\begin{pmatrix} g_{11}^* & g_{12}^* \\ g_{21}^* & g_{22}^* \end{pmatrix} \begin{pmatrix} (b_{11}^1)^* \\ (b_{21}^1)^* \end{pmatrix} = \eta_1 \begin{pmatrix} -h_{21} & -h_{22} \\ h_{11} & h_{12} \end{pmatrix} \begin{pmatrix} a_{11}^1 \\ a_{21}^1 \end{pmatrix} \tag{2.23}$$

$$\begin{pmatrix} g_{11}^* & g_{12}^* \\ g_{21}^* & g_{22}^* \end{pmatrix} \begin{pmatrix} (b_{12}^2)^* \\ (b_{22}^2)^* \end{pmatrix} = \eta_2 \begin{pmatrix} -h_{21} & -h_{22} \\ h_{11} & h_{12} \end{pmatrix} \begin{pmatrix} a_{12}^2 \\ a_{22}^2 \end{pmatrix} \tag{2.24}$$

with the normalization conditions of the precoders represented by

$$|a_{11}^1|^2 + |a_{21}^1|^2 = |b_{11}^1|^2 + |b_{21}^1|^2 = \frac{1}{2} \tag{2.25}$$

$$|a_{12}^2|^2 + |a_{22}^2|^2 = |b_{12}^2|^2 + |b_{22}^2|^2 = \frac{1}{2} \tag{2.26}$$

where we have used Eqs. (4.15) and (2.18). Note that Eqs. (2.25) and (2.26) are non-linear equations, if numerical algorithms are used to solve these equations directly, the encoding complexity will be increased exponentially with respect to the number of users and antennas. So we need to find a low-complexity method to determine the precoder parameters. First, we consider Eqs. (2.23) and (2.25).

From (2.23), we have

$$\begin{pmatrix} (b_{11}^1)^* \\ (b_{21}^1)^* \end{pmatrix} = \eta_1 \begin{pmatrix} g_{11}^* & g_{12}^* \\ g_{21}^* & g_{22}^* \end{pmatrix}^{-1} \begin{pmatrix} -h_{21} & -h_{22} \\ h_{11} & h_{12} \end{pmatrix} \begin{pmatrix} a_{11}^1 \\ a_{21}^1 \end{pmatrix} \tag{2.27}$$

Let

$$\mathbf{Q} = \begin{pmatrix} g_{11}^* & g_{12}^* \\ g_{21}^* & g_{22}^* \end{pmatrix}^{-1} \begin{pmatrix} -h_{21} & -h_{22} \\ h_{11} & h_{12} \end{pmatrix} \tag{2.28}$$

By (2.25) and (2.27), we have

$$|b_{11}^1|^2 + |b_{21}^1|^2 = \left\| \eta_1 \mathbf{Q} \begin{pmatrix} a_{11}^1 \\ a_{21}^1 \end{pmatrix} \right\|_F^2 = \frac{1}{2} \tag{2.29}$$

Now, let us consider the Singular Value Decomposition of matrix $\mathbf{Q}$, i.e.,

$$\mathbf{Q} = \mathbf{U}\boldsymbol{\Sigma}\mathbf{V}^\dagger = \mathbf{U}\mathrm{diag}(\lambda_1, \lambda_2)\mathbf{V}^\dagger \tag{2.30}$$

where $\mathbf{U}$ and $\mathbf{V}$ are unitary matrices and $\boldsymbol{\Sigma}$ is a diagonal matrix with nonnegative diagonal elements $\{\lambda_1, \lambda_2\}$ in decreasing order. Replacing (2.30) in (3.8) results in

$$\left\| \eta_1 \mathbf{U}\boldsymbol{\Sigma}\mathbf{V}^\dagger \begin{pmatrix} a_{11}^1 \\ a_{21}^1 \end{pmatrix} \right\|_F^2 = \frac{1}{2} \tag{2.31}$$

Multiplying by a unitary matrix does not change the norm of a vector, so we have

$$\left\| \eta_1 \boldsymbol{\Sigma}\mathbf{V}^\dagger \begin{pmatrix} a_{11}^1 \\ a_{21}^1 \end{pmatrix} \right\|_F^2 = \frac{1}{2} \tag{2.32}$$

Then defining

$$\begin{pmatrix} x_1 \\ x_2 \end{pmatrix} = \mathbf{V}^\dagger \begin{pmatrix} a_{11}^1 \\ a_{21}^1 \end{pmatrix} \tag{2.33}$$

and replacing it in (3.20) results in

$$\left\| \eta_1 \Sigma \begin{pmatrix} x_1 \\ x_2 \end{pmatrix} \right\|_F^2 = |\eta_1|^2 |\lambda_1|^2 |x_1|^2 + |\eta_1|^2 |\lambda_2|^2 |x_2|^2 = \frac{1}{2} \tag{2.34}$$

Since $\mathbf{V}^\dagger$ is unitary, by (2.33) and (2.25), we have

$$|x_1|^2 + |x_2|^2 = |a_{11}^1|^2 + |a_{21}^1|^2 = \frac{1}{2} \tag{2.35}$$

If we let $\overline{x}_1 = |x_1|^2$, $\overline{x}_2 = |x_2|^2$, then we can replace the nonlinear equations (2.23) and (2.25) by the following two linear equations:

$$\overline{x}_1 + \overline{x}_2 = \frac{1}{2} \tag{2.36}$$

$$|\lambda_1|^2 \overline{x}_1 + |\lambda_2|^2 \overline{x}_2 = \frac{1}{2|\eta_1|^2} \tag{2.37}$$

In the next step, we will choose the precoder parameters satisfying Eqs. (3.26) and (3.28). Note that the computational complexity of solving these linear equations is very low compared with that of solving Eqs. (2.23) and (2.25).

Step 4: Choosing the precoder parameters:

Note that in Eqs. (3.26) and (3.28), the number of unknown parameters is more than the number of equations. Therefore, the solution to achieve interference cancellation and full diversity for each user is not unique. Different solutions may lead to different coding gains and different complexity. Our emphasis in this section is on low complexity. However, in Sect. 2.4, we will show that by adding a rotation matrix, we can also maximize the coding gain. In what follows, first we choose η_1. At the first time slot, we choose $\eta_1 = \frac{1}{\lambda_1}$. Then (3.26) and (3.28) become

$$\overline{x}_1 + \overline{x}_2 = \frac{1}{2} \tag{2.38}$$

$$|\lambda_1|^2 \overline{x}_1 + |\lambda_2|^2 \overline{x}_2 = \frac{1}{2} |\lambda_1|^2 \tag{2.39}$$

It is easy to derive $\overline{x}_1 = \frac{1}{2}, \overline{x}_2 = 0$. By (2.33), we have

$$\begin{pmatrix} a_{11}^1 \\ a_{21}^1 \end{pmatrix} = \mathbf{V} \begin{pmatrix} \frac{1}{\sqrt{2}} \\ 0 \end{pmatrix} = \frac{1}{\sqrt{2}} \mathbf{V}(1) \tag{2.40}$$

Then, by (2.27), we have

$$\begin{pmatrix} (b^1_{11})^* \\ (b^1_{21})^* \end{pmatrix} = \eta_1 \mathbf{Q} \begin{pmatrix} a^1_{11} \\ a^1_{21} \end{pmatrix} = \frac{1}{\lambda_1} \mathbf{U}\boldsymbol{\Sigma}\mathbf{V}^{\dagger}\mathbf{V} \begin{pmatrix} \frac{1}{\sqrt{2}} \\ 0 \end{pmatrix} = \frac{1}{\sqrt{2}}\mathbf{U}(1) \tag{2.41}$$

Finally, by (4.17), we can determine the precoders $\mathbf{A}^1$ for User 1 and $\mathbf{B}^1$ for User 2 completely at time slot 1 as follows

$$\mathbf{A}^1 = \frac{1}{\sqrt{2}}[\mathbf{V}(1), \mathbf{V}(1)], \quad \mathbf{B}^1 = \frac{1}{\sqrt{2}}[\mathbf{U}(1), \mathbf{U}(1)]^* \tag{2.42}$$

At time slot 2, we need to solve Eqs. (2.24) and (2.26). By the same method used at time slot 1, we can arrive at

$$\overline{x}_1 + \overline{x}_2 = \frac{1}{2} \tag{2.43}$$

$$|\lambda_1|^2\overline{x}_1 + |\lambda_2|^2\overline{x}_2 = \frac{1}{2|\eta_2|^2} \tag{2.44}$$

Then we choose $\eta_2 = \frac{1}{\lambda_2}$. Replacing η_2 in (2.43) and (2.44) results in

$$\overline{x}_1 + \overline{x}_2 = \frac{1}{2} \tag{2.45}$$

$$|\lambda_1|^2\overline{x}_1 + |\lambda_2|^2\overline{x}_2 = \frac{1}{2}|\lambda_2|^2 \tag{2.46}$$

It is easy to derive $\overline{x}_1 = 0, \overline{x}_2 = \frac{1}{2}$. So we have

$$\begin{pmatrix} a^2_{12} \\ a^2_{22} \end{pmatrix} = \mathbf{V} \begin{pmatrix} 0 \\ \frac{1}{\sqrt{2}} \end{pmatrix} = \frac{1}{\sqrt{2}}\mathbf{V}(2) \tag{2.47}$$

and

$$\begin{pmatrix} (b^2_{12})^* \\ (b^2_{22})^* \end{pmatrix} = \eta_2 \mathbf{Q} \begin{pmatrix} a^2_{12} \\ a^2_{22} \end{pmatrix} = \frac{1}{\lambda_2} \mathbf{U}\boldsymbol{\Sigma}\mathbf{V}^{\dagger}\mathbf{V} \begin{pmatrix} 0 \\ \frac{1}{\sqrt{2}} \end{pmatrix} = \frac{1}{\sqrt{2}}\mathbf{U}(2) \tag{2.48}$$

Finally, by (4.17), we can determine the precoders $\mathbf{A}^2$ for User 1 and $\mathbf{B}^2$ for User 2 completely at time slot 2 as follows

$$\mathbf{A}^2 = \frac{1}{\sqrt{2}}[\mathbf{V}(2), \mathbf{V}(2)], \quad \mathbf{B}^2 = \frac{1}{\sqrt{2}}[\mathbf{U}(2), \mathbf{U}(2)]^* \tag{2.49}$$

So far, we have designed the precoders for both users through the above 4 steps when the channel information is known at the transmitter.

2.3 Decoding

In this section, we focus on the decoding. We start with Eq. (2.20). Note that (2.20) can also be written as

$$
\begin{pmatrix} y_1^1 \\ (y_1^2)^* \\ y_2^1 \\ (y_2^2)^* \end{pmatrix} = \sqrt{E_s}\, \begin{pmatrix} \widehat{h}_{11}^1 & \widehat{h}_{11}^1 & \widehat{g}_{11}^1 & \widehat{g}_{11}^1 \\ (\widehat{h}_{12}^2)^* & -(\widehat{h}_{12}^2)^* & (\widehat{g}_{12}^2)^* & -(\widehat{g}_{12}^2)^* \\ \widehat{h}_{21}^1 & \widehat{h}_{21}^1 & \widehat{g}_{21}^1 & \widehat{g}_{21}^1 \\ (\widehat{h}_{22}^2)^* & -(\widehat{h}_{22}^2)^* & (\widehat{g}_{22}^2)^* & -(\widehat{g}_{22}^2)^* \end{pmatrix} \begin{pmatrix} c_1 \\ c_2 \\ s_1 \\ s_2 \end{pmatrix} + \begin{pmatrix} n_1^1 \\ (n_1^2)^* \\ n_2^1 \\ (n_2^2)^* \end{pmatrix}
\tag{2.50}
$$

and we define

$$
\overline{\mathbf{H}} = \begin{pmatrix} \overline{\mathbf{H}}_1 & \overline{\mathbf{G}}_1 \\ \overline{\mathbf{H}}_2 & \overline{\mathbf{G}}_2 \end{pmatrix} = \begin{pmatrix} \widehat{h}_{11}^1 & \widehat{h}_{11}^1 & \widehat{g}_{11}^1 & \widehat{g}_{11}^1 \\ (\widehat{h}_{12}^2)^* & -(\widehat{h}_{12}^2)^* & (\widehat{g}_{12}^2)^* & -(\widehat{g}_{12}^2)^* \\ \widehat{h}_{21}^1 & \widehat{h}_{21}^1 & \widehat{g}_{21}^1 & \widehat{g}_{21}^1 \\ (\widehat{h}_{22}^2)^* & -(\widehat{h}_{22}^2)^* & (\widehat{g}_{22}^2)^* & -(\widehat{g}_{22}^2)^* \end{pmatrix}, \quad \overline{\mathbf{n}} = \begin{pmatrix} n_1^1 \\ (n_1^2)^* \\ n_2^1 \\ (n_2^2)^* \end{pmatrix}
\tag{2.51}
$$

where

$$
\overline{\mathbf{H}}_1 = \begin{pmatrix} \widehat{h}_{11}^1 & \widehat{h}_{11}^1 \\ (\widehat{h}_{12}^2)^* & -(\widehat{h}_{12}^2)^* \end{pmatrix}, \quad \overline{\mathbf{G}}_1 = \begin{pmatrix} \widehat{g}_{11}^1 & \widehat{g}_{11}^1 \\ (\widehat{g}_{12}^2)^* & -(\widehat{g}_{12}^2)^* \end{pmatrix}
$$

$$
\overline{\mathbf{H}}_2 = \begin{pmatrix} \widehat{h}_{21}^1 & \widehat{h}_{21}^1 \\ (\widehat{h}_{22}^2)^* & -(\widehat{h}_{22}^2)^* \end{pmatrix}, \quad \overline{\mathbf{G}}_2 = \begin{pmatrix} \widehat{g}_{21}^1 & \widehat{g}_{21}^1 \\ (\widehat{g}_{22}^2)^* & -(\widehat{g}_{22}^2)^* \end{pmatrix}
\tag{2.52}
$$

Note that $\overline{\mathbf{H}}$ has a quasi-orthogonal structure, i.e., the first two columns are orthogonal to the second two columns. If we multiply both sides of Eq. (2.50) with $\overline{\mathbf{H}}^\dagger$, we will have

$$
\overline{\mathbf{H}}^\dagger \begin{pmatrix} y_1^1 \\ (y_1^2)^* \\ y_2^1 \\ (y_2^2)^* \end{pmatrix} = \sqrt{E_s}\, \begin{pmatrix} \overline{\mathbf{H}}_1^\dagger \overline{\mathbf{H}}_1 + \overline{\mathbf{H}}_2^\dagger \overline{\mathbf{H}}_2 & 0 \\ 0 & \overline{\mathbf{G}}_1^\dagger \overline{\mathbf{G}}_1 + \overline{\mathbf{G}}_2^\dagger \overline{\mathbf{G}}_2 \end{pmatrix} \begin{pmatrix} c_1 \\ c_2 \\ s_1 \\ s_2 \end{pmatrix} + \overline{\mathbf{H}}^\dagger \overline{\mathbf{n}}
\tag{2.53}
$$

Now we define

$$
\widetilde{\mathbf{y}} = \begin{pmatrix} \widetilde{\mathbf{y}}_1 \\ \widetilde{\mathbf{y}}_2 \end{pmatrix} = \overline{\mathbf{H}}^\dagger \begin{pmatrix} y_1^1 \\ (y_1^2)^* \\ y_2^1 \\ (y_2^2)^* \end{pmatrix}
\tag{2.54}
$$

where $\widetilde{\mathbf{y}}_1 = \begin{pmatrix} \widetilde{y}(1,1) \\ \widetilde{y}(2,1) \end{pmatrix}, \widetilde{\mathbf{y}}_2 = \begin{pmatrix} \widetilde{y}(3,1) \\ \widetilde{y}(4,1) \end{pmatrix}$. Note that the noise elements of $\overline{\mathbf{H}}^\dagger \overline{\mathbf{n}}$ are correlated with covariance matrix $\overline{\mathbf{H}}^\dagger \overline{\mathbf{H}}$. We can whiten this noise vector by multiplying both sides of (2.54) by the matrix $(\overline{\mathbf{H}}^\dagger \overline{\mathbf{H}})^{-\frac{1}{2}}$ as follows

$$(\overline{\mathbf{H}}^{\dagger}\overline{\mathbf{H}})^{-\frac{1}{2}}\widetilde{\mathbf{y}} = \sqrt{E_s}(\overline{\mathbf{H}}^{\dagger}\overline{\mathbf{H}})^{\frac{1}{2}}\begin{pmatrix} c_1 \\ c_2 \\ s_1 \\ s_2 \end{pmatrix} + \widehat{\mathbf{n}} \tag{2.55}$$

where $\widehat{\mathbf{n}} = (\overline{\mathbf{H}}^{\dagger}\overline{\mathbf{H}})^{-\frac{1}{2}}(\overline{\mathbf{H}}^{\dagger}\overline{\mathbf{n}})$ has uncorrelated elements $\sim CN(0, 1)$. If we define

$$\widehat{\mathbf{H}} = \overline{\mathbf{H}}_1^{\dagger}\overline{\mathbf{H}}_1 + \overline{\mathbf{H}}_2^{\dagger}\overline{\mathbf{H}}_2 \tag{2.56}$$

$$\widehat{\mathbf{G}} = \overline{\mathbf{G}}_1^{\dagger}\overline{\mathbf{G}}_1 + \overline{\mathbf{G}}_2^{\dagger}\overline{\mathbf{G}}_2 \tag{2.57}$$

$$\widehat{\mathbf{n}} = \begin{pmatrix} \widehat{\mathbf{n}}_1 \\ \widehat{\mathbf{n}}_2 \end{pmatrix}, \widehat{\mathbf{n}}_1 = \begin{pmatrix} \widehat{n}(1, 1) \\ \widehat{n}(2, 1) \end{pmatrix}, \widehat{\mathbf{n}}_2 = \begin{pmatrix} \widehat{n}(3, 1) \\ \widehat{n}(4, 1) \end{pmatrix} \tag{2.58}$$

Then (2.55) is equivalent to the following two equations

$$\widehat{\mathbf{H}}^{-\frac{1}{2}}\widetilde{\mathbf{y}}_1 = \sqrt{E_s}\widehat{\mathbf{H}}^{\frac{1}{2}}\begin{pmatrix} c_1 \\ c_2 \end{pmatrix} + \widehat{\mathbf{n}}_1 \tag{2.59}$$

$$\widehat{\mathbf{G}}^{-\frac{1}{2}}\widetilde{\mathbf{y}}_2 = \sqrt{E_s}\widehat{\mathbf{G}}^{\frac{1}{2}}\begin{pmatrix} s_1 \\ s_2 \end{pmatrix} + \widehat{\mathbf{n}}_2 \tag{2.60}$$

So we can realize interference cancellation and pairwise complex symbol decoding for each user. If we use real symbols, instead of complex symbols, we can achieve symbol-by-symbol decoding using orthogonal designs instead of quasi-orthogonal designs. In other words, we can design precoders such that all columns of the equivalent matrix $\widetilde{\mathbf{H}}$ in Eq. (2.14) are orthogonal to each other.

When QAM is adopted, we show that we can further reduce the decoding complexity as follows. Note that for 2×2 complex matrix $\mathbf{Z} = \begin{pmatrix} \alpha & \alpha \\ \beta & -\beta \end{pmatrix}$, $\mathbf{Z}^{\dagger}\mathbf{Z} = \begin{pmatrix} |\alpha|^2 + |\beta|^2 & |\alpha|^2 - |\beta|^2 \\ |\alpha|^2 - |\beta|^2 & |\alpha|^2 + |\beta|^2 \end{pmatrix}$, which is a real matrix. So matrices $\widehat{\mathbf{H}}$ and $\widehat{\mathbf{G}}$ in (2.59), (2.60) are all real matrices. Then (2.59), (2.60) are equivalent to the following four equations

$$\widehat{\mathbf{H}}^{-\frac{1}{2}}Real\{\widetilde{\mathbf{y}}_1\} = \sqrt{E_s}\widehat{\mathbf{H}}^{\frac{1}{2}}\begin{pmatrix} c_{1R} \\ c_{2R} \end{pmatrix} + Real\{\widehat{\mathbf{n}}_1\} \tag{2.61}$$

$$\widehat{\mathbf{H}}^{-\frac{1}{2}}Imag\{\widetilde{\mathbf{y}}_1\} = \sqrt{E_s}\widehat{\mathbf{H}}^{\frac{1}{2}}\begin{pmatrix} c_{1I} \\ c_{2I} \end{pmatrix} + Imag\{\widehat{\mathbf{n}}_1\} \tag{2.62}$$

$$\widehat{\mathbf{G}}^{-\frac{1}{2}}Real\{\widetilde{\mathbf{y}}_2\} = \sqrt{E_s}\widehat{\mathbf{G}}^{\frac{1}{2}}\begin{pmatrix} s_{1R} \\ s_{2R} \end{pmatrix} + Real\{\widehat{\mathbf{n}}_2\} \tag{2.63}$$

$$\widehat{\mathbf{G}}^{-\frac{1}{2}} Imag\{\widetilde{\mathbf{y}}_2\} = \sqrt{E_s}\widehat{\mathbf{G}}^{\frac{1}{2}} \begin{pmatrix} s_{1I} \\ s_{2I} \end{pmatrix} + Imag\{\widehat{\mathbf{n}}_2\} \tag{2.64}$$

where $Real\{\mathbf{z}\}$, $Imag\{\mathbf{z}\}$ denote the real and imaginary parts of vector $\mathbf{z}$, respectively. So we can use the Maximum-Likelihood method to detect (c_{1R}, c_{2R}), (c_{1I}, c_{2I}), (s_{1R}, s_{2R}), (s_{1I}, s_{2I}) separately. For example, by (2.61), we can detect (c_{1R}, c_{2R}) by

$$\widehat{c}_{1R}, \widehat{c}_{2R} = \arg \min_{c_{1R}, c_{2R}} \left\| \widehat{\mathbf{H}}^{-\frac{1}{2}} Real\{\widetilde{\mathbf{y}}_1\} - \sqrt{E_s}\widehat{\mathbf{H}}^{\frac{1}{2}} \begin{pmatrix} c_{1R} \\ c_{2R} \end{pmatrix} \right\|_F^2 \tag{2.65}$$

Similarly, using (2.62–2.64), we can detect all other codewords.

2.4 Proof of Full Diversity

Diversity is usually defined as the exponent of Signal-to-Noise-Ratio (SNR) in the error rate expression at high-SNR. Mathematically, the diversity order can be defined as

$$d = - \lim_{\rho \to \infty} \frac{\log P_e}{\log \rho} \tag{2.66}$$

where ρ denotes the SNR and P_e represents the probability of error. We first consider (2.59) to analyze the diversity for User 1. Here we add a unitary rotation $\mathbf{R}$ to $\begin{pmatrix} c_1 \\ c_2 \end{pmatrix}$. Thus, the data vector $\mathbf{d} = \mathbf{R} \begin{pmatrix} c_1 \\ c_2 \end{pmatrix}$ and we define the error matrix $\boldsymbol{\varepsilon} = \begin{pmatrix} c_1 \\ c_2 \end{pmatrix} - \begin{pmatrix} \widehat{c}_1 \\ \widehat{c}_2 \end{pmatrix}$. By (2.59), the pairwise error probability (PEP) can be given by the Gaussian tail function as [2]

$$P(\mathbf{d} \to \overline{\mathbf{d}}|\widehat{\mathbf{H}}) = Q \left(\sqrt{\frac{\rho \|\widehat{\mathbf{H}}^{\frac{1}{2}} \mathbf{R}\boldsymbol{\varepsilon}\|_F^2}{4}} \right) \tag{2.67}$$

Now we assume $\overline{\mathbf{H}}_1$ and $\overline{\mathbf{H}}_2$ have the following singular value decompositions

$$\overline{\mathbf{H}}_1 = \mathbf{U}_1 \Lambda_1 \mathbf{V}_1 = \mathbf{U}_1 \text{diag}\{\lambda_{11}, \lambda_{12}\}\mathbf{V}_1 \tag{2.68}$$

$$\overline{\mathbf{H}}_2 = \mathbf{U}_2 \Lambda_2 \mathbf{V}_2 = \mathbf{U}_2 \text{diag}\{\lambda_{21}, \lambda_{22}\}\mathbf{V}_2 \tag{2.69}$$

Since $\overline{\mathbf{H}}_1^\dagger \overline{\mathbf{H}}_1 = \mathbf{V}_1^\dagger \Lambda_1^\dagger \Lambda_1 \mathbf{V}_1$ and $\overline{\mathbf{H}}_2^\dagger \overline{\mathbf{H}}_2 = \mathbf{V}_2^\dagger \Lambda_2^\dagger \Lambda_2 \mathbf{V}_2$ are both block-circulant matrices, $\mathbf{V}_1 = \mathbf{V}_2 = \frac{1}{\sqrt{2}} \begin{pmatrix} 1 & 1 \\ 1 & -1 \end{pmatrix}$ [3]. We let $\widetilde{\mathbf{V}}_1 = \widetilde{\mathbf{V}}_2 = \begin{pmatrix} 1 & 1 \\ 1 & -1 \end{pmatrix}$ and $\widetilde{\Lambda}_1 = \frac{1}{\sqrt{2}}\Lambda_1 = \text{diag}\{\widetilde{\lambda}_{11}, \widetilde{\lambda}_{12}\}$, $\widetilde{\Lambda}_2 = \frac{1}{\sqrt{2}}\Lambda_2 = \text{diag}\{\widetilde{\lambda}_{21}, \widetilde{\lambda}_{22}\}$. Therefore, (2.67) can be written as

$$P(\mathbf{d} \to \overline{\mathbf{d}}|\widehat{\mathbf{H}}) = Q\left(\sqrt{\frac{\rho[\boldsymbol{\varepsilon}^\dagger \mathbf{R}^\dagger \widetilde{\mathbf{V}}_1^\dagger (\widetilde{\Lambda}_1^\dagger \widetilde{\Lambda}_1 + \widetilde{\Lambda}_2^\dagger \widetilde{\Lambda}_2)\widetilde{\mathbf{V}}_1 \mathbf{R}\boldsymbol{\varepsilon}]}{4}}\right) \tag{2.70}$$

By replacing $\Phi = \widetilde{\mathbf{V}}_1 \mathbf{R}\boldsymbol{\varepsilon}$ in (2.70), we have

$$P(\mathbf{d} \to \overline{\mathbf{d}}|\widehat{\mathbf{H}}) = Q\left(\sqrt{\frac{\rho \sum_{i=1}^2 \sum_{j=1}^2 |\Phi(j,1)|^2 |\widetilde{\lambda}_{i,j}|^2}{4}}\right) \tag{2.71}$$

Using the inequality $Q(x) \le \exp(-\frac{x^2}{2})$ results in

$$P(\mathbf{d} \to \overline{\mathbf{d}}|\widehat{\mathbf{H}}) \le \exp\left(-\frac{\rho \sum_{i=1}^2 \sum_{j=1}^2 |\Phi(j,1)|^2 |\widetilde{\lambda}_{i,j}|^2}{8}\right) \tag{2.72}$$

Now we evaluate the distribution of $\widetilde{\lambda}_{i,j}$. We know that

$$\overline{\mathbf{H}}_1 = \mathbf{U}_1 \begin{pmatrix} \widetilde{\lambda}_{11} & 0 \\ 0 & \widetilde{\lambda}_{12} \end{pmatrix} \begin{pmatrix} 1 & 1 \\ 1 & -1 \end{pmatrix}, \quad \overline{\mathbf{H}}_2 = \mathbf{U}_2 \begin{pmatrix} \widetilde{\lambda}_{21} & 0 \\ 0 & \widetilde{\lambda}_{22} \end{pmatrix} \begin{pmatrix} 1 & 1 \\ 1 & -1 \end{pmatrix} \tag{2.73}$$

Therefore, $\begin{pmatrix} \widetilde{\lambda}_{11} \\ \widetilde{\lambda}_{12} \end{pmatrix} = \mathbf{U}_1^\dagger \overline{\mathbf{H}}_1(1)$ and $\begin{pmatrix} \widetilde{\lambda}_{21} \\ \widetilde{\lambda}_{22} \end{pmatrix} = \mathbf{U}_2^\dagger \overline{\mathbf{H}}_2(1)$. By (2.52), (2.8), and (2.10), we know that conditioned on $\mathbf{V}$, each element of $\overline{\mathbf{H}}_1(1)$ and $\overline{\mathbf{H}}_2(1)$ will be i.i.d complex Gaussian random variables with mean 0 and variance 1. Multiplying by unitary matrices $\mathbf{U}_1^\dagger$ and $\mathbf{U}_2^\dagger$ does not change the distribution. So $\widetilde{\lambda}_{11}$, $\widetilde{\lambda}_{12}$, $\widetilde{\lambda}_{21}$, $\widetilde{\lambda}_{22}$ are all i.i.d complex Gaussian random variables with mean 0 and variance 1. Their magnitudes, $|\widetilde{\lambda}_{i,j}|^2$, are Rayleigh with the probability density function $f(|\widetilde{\lambda}_{i,j}|) = 2|\widetilde{\lambda}_{i,j}| \exp(-|\widetilde{\lambda}_{i,j}|^2)$. Using the distribution of $|\widetilde{\lambda}_{i,j}|$, we have

$$P(\mathbf{d} \to \overline{\mathbf{d}}) = E[P(\mathbf{d} \to \overline{\mathbf{d}}|\widehat{\mathbf{H}})] = E_{\mathbf{V}}[E_{\widehat{\mathbf{H}}}[P(\mathbf{d} \to \overline{\mathbf{d}}|\widehat{\mathbf{H}})]|\mathbf{V}]$$

$$\le E_{\mathbf{V}}\left[E_{\widehat{\mathbf{H}}}\left[\exp\left(-\frac{\rho \sum_{i=1}^2 \sum_{j=1}^2 |\Phi(j,1)|^2 |\widetilde{\lambda}_{i,j}|^2}{8}\right)\right]\bigg|\mathbf{V}\right]$$

$$= E_{\mathbf{V}}\left[\frac{1}{\prod_{j=1}^2 [1 + (\rho|\Phi(j,1)|^2/8)]^2}\bigg|\mathbf{V}\right]$$

$$= \frac{1}{\prod_{j=1}^2 [1 + (\rho|\Phi(j,1)|^2/8)]^2} \tag{2.74}$$

At high SNRs, one can neglect the one in the denominator and get

$$P(\mathbf{d} \to \bar{\mathbf{d}}) \leq \left(\frac{\rho}{8}\right)^{-4} \prod_{j-1}^{2} |\Phi(j, 1)|^{-4} \tag{2.75}$$

By (5.50), it is easy to show that the diversity is 4 if we choose a proper unitary rotation matrix $\mathbf{R}$ such that $\prod_{j=1}^{2} |\Phi(j, 1)| \neq 0$. The best known rotations for QAM to maximize the minimum product distance are provided in [4]. Similarly, we can prove that the diversity for User 2 is also 4. Therefore, our scheme can achieve full diversity for each user. Similarly, it can be shown that the system provides full diversity when we use Eqs. (2.61–2.64) to simplify the decoding complexity for QAM.

2.5 Extension to Two Users with More than Two Transmit Antennas

In this section, we show that the scheme used for 2 users each with 2 transmit antennas can also be extended to 2 users each with more than 2 transmit antennas. Assume we have 2 users each with $N = 2n$ transmit antennas. At the first N time slots, Users 1 and 2 send codewords

$$\mathbf{C} = \begin{pmatrix} -c_1 & c_1 & \cdots & c_1 \\ c_2 & -c_2 & \cdots & c_2 \\ \vdots & \vdots & \ddots & \vdots \\ c_N & c_N & \cdots & -c_N \end{pmatrix}, \quad \mathbf{S} = \begin{pmatrix} -s_1 & s_1 & \cdots & s_1 \\ s_2 & -s_2 & \cdots & s_2 \\ \vdots & \vdots & \ddots & \vdots \\ s_N & s_N & \cdots & -s_N \end{pmatrix} \tag{2.76}$$

respectively. The received signals at time slot i, $i = 1, \ldots, N$, is denoted by

$$\mathbf{y}^i = \begin{pmatrix} y_1^i \\ y_2^i \end{pmatrix} \tag{2.77}$$

Within these N time slots, the channel matrices for Users 1 and 2 are

$$\mathbf{H} = \begin{pmatrix} h_{11} & h_{12} & \cdots & h_{1N} \\ h_{21} & h_{22} & \cdots & h_{2N} \end{pmatrix}, \quad \mathbf{G} = \begin{pmatrix} g_{11} & g_{12} & \cdots & g_{1N} \\ g_{21} & g_{22} & \cdots & g_{2N} \end{pmatrix} \tag{2.78}$$

respectively. At time slot i, $i = 1, \ldots, N$, the precoders for Users 1 and 2 are

$$\mathbf{A}^i = \begin{pmatrix} a_{11}^i & a_{12}^i & \cdots & a_{1N}^i \\ a_{21}^i & a_{22}^i & \cdots & a_{2N}^i \\ \vdots & \vdots & \ddots & \vdots \\ a_{N1}^i & a_{N2}^i & \cdots & a_{NN}^i \end{pmatrix}, \quad \mathbf{B}^i = \begin{pmatrix} b_{11}^i & b_{12}^i & \cdots & b_{1N}^i \\ b_{21}^i & b_{22}^i & \cdots & b_{2N}^i \\ \vdots & \vdots & \ddots & \vdots \\ b_{N1}^i & b_{N2}^i & \cdots & b_{NN}^i \end{pmatrix} \tag{2.79}$$

respectively. We follow the steps in Sect. 2.1 to design the precoders.

Step 1: Deriving the equivalent channel equations:

At time slot i, the signal model can be written as

$$\mathbf{y}^i = \sqrt{E_s}\mathbf{H}\mathbf{A}^i\mathbf{C}(i) + \sqrt{E_s}\mathbf{G}\mathbf{B}^i\mathbf{S}(i) + \mathbf{W}^i$$
$$= \sqrt{E_s}\widehat{\mathbf{H}}^i\mathbf{C}(i) + \sqrt{E_s}\widehat{\mathbf{G}}^i\mathbf{S}(i) + \mathbf{W}^i \tag{2.80}$$

where $\widehat{\mathbf{H}}^i$ and $\widehat{\mathbf{G}}^i$ denote the equivalent channel matrices for Users 1 and 2 at time slot i, respectively. Combining channel equations at the first N time slots, we have

$$\begin{pmatrix} y_1^1 \\ y_2^1 \\ y_1^2 \\ y_2^2 \\ \vdots \\ y_1^N \\ y_2^N \end{pmatrix} = \sqrt{E_s} \begin{pmatrix} -\widehat{\mathbf{H}}^1(1) & \widehat{\mathbf{H}}^1(2) & \cdots & \widehat{\mathbf{H}}^1(N) & -\widehat{\mathbf{G}}^1(1) & \widehat{\mathbf{G}}^1(2) & \cdots & \widehat{\mathbf{G}}^1(N) \\ \widehat{\mathbf{H}}^2(1) & -\widehat{\mathbf{H}}^2(2) & \cdots & \widehat{\mathbf{H}}^2(N) & \widehat{\mathbf{G}}^2(1) & -\widehat{\mathbf{G}}^2(2) & \cdots & \widehat{\mathbf{G}}^2(N) \\ \vdots & \vdots & \ddots & \vdots & \vdots & \vdots & \ddots & \vdots \\ \widehat{\mathbf{H}}^N(1) & \widehat{\mathbf{H}}^N(2) & \cdots & -\widehat{\mathbf{H}}^N(N) & \widehat{\mathbf{G}}^N(1) & \widehat{\mathbf{G}}^N(2) & \cdots & -\widehat{\mathbf{G}}^N(N) \end{pmatrix}$$

$$\times \begin{pmatrix} c_1 \\ c_2 \\ \vdots \\ c_N \\ s_1 \\ s_2 \\ \vdots \\ s_N \end{pmatrix} + \begin{pmatrix} n_1^1 \\ n_2^1 \\ n_1^2 \\ n_2^2 \\ \vdots \\ n_1^N \\ n_2^N \end{pmatrix} \tag{2.81}$$

Here we let

$$\widetilde{\mathbf{H}} = \begin{pmatrix} -\widehat{\mathbf{H}}^1(1) & \widehat{\mathbf{H}}^1(2) & \cdots & \widehat{\mathbf{H}}^1(N) & -\widehat{\mathbf{G}}^1(1) & \widehat{\mathbf{G}}^1(2) & \cdots & \widehat{\mathbf{G}}^1(N) \\ \widehat{\mathbf{H}}^2(1) & -\widehat{\mathbf{H}}^2(2) & \cdots & \widehat{\mathbf{H}}^2(N) & \widehat{\mathbf{G}}^2(1) & -\widehat{\mathbf{G}}^2(2) & \cdots & \widehat{\mathbf{G}}^2(N) \\ \vdots & \vdots & \ddots & \vdots & \vdots & \vdots & \ddots & \vdots \\ \widehat{\mathbf{H}}^N(1) & \widehat{\mathbf{H}}^N(2) & \cdots & -\widehat{\mathbf{H}}^N(N) & \widehat{\mathbf{G}}^N(1) & \widehat{\mathbf{G}}^N(2) & \cdots & -\widehat{\mathbf{G}}^N(N) \end{pmatrix} \tag{2.82}$$

Step 2: Creating the orthogonal structure of signal vectors:

Let

$$\mathbf{A}^i(1) = \mathbf{A}^i(2) = \mathbf{A}^i(3) = \cdots = \mathbf{A}^i(N) \tag{2.83}$$
$$\mathbf{B}^i(1) = \mathbf{B}^i(2) = \mathbf{B}^i(3) = \cdots = \mathbf{B}^i(N) \tag{2.84}$$

Equations (2.83) and (2.84) will result in

$$\widehat{\mathbf{H}}^i(1) = \widehat{\mathbf{H}}^i(2) = \cdots = \widehat{\mathbf{H}}^i(N) \tag{2.85}$$

$$\widehat{\mathbf{G}}^i(1) = \widehat{\mathbf{G}}^i(2) = \cdots = \widehat{\mathbf{G}}^i(N) \tag{2.86}$$

respectively. In order to make the symbols of Users 1 and 2 transmitted in two orthogonal subspaces, i.e., the first N columns of $\widetilde{\mathbf{H}}$ are orthogonal to the second N columns of $\widetilde{\mathbf{H}}$, we let

$$\begin{pmatrix} \widehat{G}^i(1,1) \\ \widehat{G}^i(2,1) \end{pmatrix} = \eta_i \begin{pmatrix} -\widehat{H}^i(2,1) \\ \widehat{H}^i(1,1) \end{pmatrix}^* \tag{2.87}$$

Step 3: Designing low-complexity algorithms to calculate the parameters of the precoders:

From (2.87), we have

$$\begin{pmatrix} g_{11} & g_{12} & \cdots & g_{1N} \\ g_{21} & g_{22} & \cdots & g_{2N} \end{pmatrix}^* \begin{pmatrix} b^i_{11} \\ b^i_{21} \\ \vdots \\ b^i_{N1} \end{pmatrix}^* = \eta_i \begin{pmatrix} -h_{21} & -h_{22} & \cdots & -h_{2N} \\ h_{11} & h_{12} & \cdots & h_{1N} \end{pmatrix} \begin{pmatrix} a^i_{11} \\ a^i_{21} \\ \vdots \\ a^i_{N1} \end{pmatrix} \tag{2.88}$$

with normalization equations

$$|a^i_{11}|^2 + |a^i_{21}|^2 + \cdots + |a^i_{N1}|^2 = \frac{1}{N}$$

$$|b^i_{11}|^2 + |b^i_{21}|^2 + \cdots + |b^i_{N1}|^2 = \frac{1}{N} \tag{2.89}$$

Note that the channel matrices in (2.88) are not square matrices. Therefore, we cannot use the reverse matrix directly as we did for the users with 2 transmit antennas in Sect. 2.1. Instead, in order to simplify the precoder design, at the first 2 time slots, we let all the elements in complex vector

$$\mathbf{a}^i = \begin{pmatrix} a^i_{11} & a^i_{21} & \cdots & a^i_{N1} \end{pmatrix}^T, \quad i = 1, 2 \tag{2.90}$$

be zero except for the first 2 elements and also let all the elements in

$$\mathbf{b}^i = \begin{pmatrix} b^i_{11} & b^i_{21} & \cdots & b^i_{N1} \end{pmatrix}^T, \quad i = 1, 2 \tag{2.91}$$

be zero except for the first 2 elements. By the above choices for $\mathbf{a}^i$ and $\mathbf{b}^i$, Eq. (2.88) results in

$$\begin{pmatrix} g_{11} & g_{12} \\ g_{21} & g_{22} \end{pmatrix}^* \begin{pmatrix} b^i_{11} \\ b^i_{21} \end{pmatrix}^* = \eta_i \begin{pmatrix} -h_{21} & -h_{22} \\ h_{11} & h_{12} \end{pmatrix} \begin{pmatrix} a^i_{11} \\ a^i_{21} \end{pmatrix} \tag{2.92}$$

which is exactly the same as (2.27). Following the steps in Sect. 2.1, Eqs. (2.89) and (2.92) result in

$$|x_1|^2 + |x_2|^2 = \frac{1}{N}$$

$$|\lambda_1|^2|x_1|^2 + |\lambda_2|^2|x_2|^2 = \frac{1}{N} \cdot \frac{1}{|\eta_i|^2} \tag{2.93}$$

Step 4: Choosing the precoder parameters:
At time slot 1, we choose $\eta_1 = \frac{1}{\lambda_1}$. It is easy to show

$$\begin{pmatrix} x_1 \\ x_2 \end{pmatrix} = \frac{1}{\sqrt{N}} \begin{pmatrix} 1 \\ 0 \end{pmatrix}, \quad \begin{pmatrix} a_{11}^1 \\ a_{21}^1 \end{pmatrix} = \frac{1}{\sqrt{N}} \mathbf{V}(1) \tag{2.94}$$

At time slot 2, we choose $\eta_2 = \frac{1}{\lambda_2}$ that results in

$$\begin{pmatrix} x_1 \\ x_2 \end{pmatrix} = \frac{1}{\sqrt{N}} \begin{pmatrix} 0 \\ 1 \end{pmatrix}, \quad \begin{pmatrix} a_{11}^2 \\ a_{21}^2 \end{pmatrix} = \frac{1}{\sqrt{N}} \mathbf{V}(2) \tag{2.95}$$

where $\mathbf{V}$ comes from the singular value decomposition in (2.30).

At time slots 3 and 4, the precoder design procedures are nearly the same as that of the first 2 time slots. The only difference is that, we let all the elements be zero except the second 2 elements in both $\mathbf{a}^i$ and $\mathbf{b}^i$, $i = 3, 4$, in order to get a square matrix like that in (2.92). Then we follow the same steps to determine the precoders at time slots 3 and 4. We repeat the same process, by shifting the window of 2 nonzero elements, until all precoders at all time slots are designed. This completes our extension to more than 2 transmit antennas. For the sake of brevity, we do not include the decoding and the proof of full diversity. They are similar in nature to what we presented earlier for users with 2 transmit antennas.

2.6 Extension to More than 2 Receive Antennas

So far, we have proposed a scheme for 2 users each with N transmit antennas and one receiver with 2 receive antennas. In this section, we consider the case of $M > 2$ receive antennas. First, note that if $M = 2m$ and $N = n \cdot M$, where m, n are positive integers, our approach in Sect. 2.5 will still work if we adjust the dimension of the transmitted signals, the received signals, and the channel matrices.

Second, for other cases, we show that our scheme combined with antenna selection can also achieve interference cancellation and full diversity for each user. In other words, extra antennas will provide extra diversity and the resulting diversity of the system is NM.

For the sake of simplicity, we consider 2 users each with 2 transmit antennas and one receiver with 3 receive antennas. The approach for a general case of N transmit and M receive antennas is similar. Our approach is to select 2 of the 3 receive antennas and use the scheme in Sect. 2.1 for the selected antennas. Now we will present our selection criterion. Note that by using the scheme proposed in Sect. 2.1, as shown in (2.72), the term that determines diversity is $\sum_{i=1}^{2} \sum_{j=1}^{2} |\Phi(j,1)|^2 |\widetilde{\lambda}_{i,j}|^2$. We know $\Phi = \widetilde{\mathbf{V}}_1 \mathbf{R}\boldsymbol{\varepsilon} = [\Phi(1,1), \Phi(2,1)]^T$ where $\widetilde{\mathbf{V}}_1$ is constant and $\boldsymbol{\varepsilon}$ is the error matrix. For a given constellation, the unitary rotation matrix $\mathbf{R}$ is chosen optimally and is fixed. So we can always find $\phi_1 = \min_{\forall d^i, d^j} |\Phi(1,1)|, i \neq j$ and $\phi_2 = \min_{\forall d^i, d^j} |\Phi(2,1)|, i \neq j$. Now we define $\varphi = \sum_{i=1}^{2} \sum_{j=1}^{2} |\phi_j|^2 |\widetilde{\lambda}_{i,j}|^2$. Different choice of receive antennas will lead to different $\widetilde{\lambda}_{i,j}$ and thus different φ. To pick 2 out of 3 antennas, we have 3 choices. We call the scenario that receive antennas 1 and 2 are chosen Case 1, the scenario that receive antennas 1 and 3 are chosen Case 2, and, the scenario that receive antennas 2 and 3 are chosen Case 3. The corresponding φ for each case is given by $\varphi_k = \sum_{i=1}^{2} \sum_{j=1}^{2} |\phi_j|^2 |\widetilde{\lambda}_{i,j}^k|^2$, $k = 1, 2, 3$. Our selection criterion is to pick the two receive antennas of Case i whose corresponding φ_i is the largest among all the three cases. In other words, if $\varphi_i = \max\{\varphi_1, \varphi_2, \varphi_3\}$, then we choose the two antennas corresponding to Case i. Obviously, by this method, we can achieve interference cancellation for each user. In what follows, we prove that we can also achieve full diversity for each user.

We first present the proof for User 1. Let us assume the channel for User 1 is $\mathbf{H} = \begin{pmatrix} h_{11} & h_{12} \\ h_{21} & h_{22} \\ h_{31} & h_{32} \end{pmatrix}$. The channels for User 1 in Cases 1, 2, 3 are $\mathbf{H}_1 = \begin{pmatrix} h_{11} & h_{12} \\ h_{21} & h_{22} \end{pmatrix}$, $\mathbf{H}_2 = \begin{pmatrix} h_{11} & h_{12} \\ h_{31} & h_{32} \end{pmatrix}$, and $\mathbf{H}_3 = \begin{pmatrix} h_{21} & h_{22} \\ h_{31} & h_{32} \end{pmatrix}$, respectively. Without loss of generality, let us assume $i = \arg\max\{\varphi_1, \varphi_2, \varphi_3\} \in \{1, 2, 3\}$ and the two receive antennas in case i is selected. By our selection criterion, we know that

$$\frac{\varphi_1 + \varphi_2 + \varphi_3}{3} \leq \varphi_i \leq \varphi_1 + \varphi_2 + \varphi_3 \tag{2.96}$$

where

$$\varphi_i = |\phi_1|^2 (|\widetilde{\lambda}_{11}^i|^2 + |\widetilde{\lambda}_{21}^i|^2) + |\phi_2|^2 (|\widetilde{\lambda}_{12}^i|^2 + |\widetilde{\lambda}_{22}^i|^2) \tag{2.97}$$

Now, let us define

$$\begin{aligned} \delta_1 = {} & |\phi_1|^2 (|\widetilde{\lambda}_{11}^1|^2 + |\widetilde{\lambda}_{21}^1|^2 + |\widetilde{\lambda}_{21}^2|^2) \\ & + |\phi_2|^2 (|\widetilde{\lambda}_{12}^1|^2 + |\widetilde{\lambda}_{22}^1|^2 + |\widetilde{\lambda}_{22}^2|^2) \end{aligned} \tag{2.98}$$

$$\begin{aligned} \delta_2 = {} & |\phi_1|^2 (|\widetilde{\lambda}_{11}^3|^2 + |\widetilde{\lambda}_{21}^3|^2 + |\widetilde{\lambda}_{11}^2|^2) \\ & + |\phi_2|^2 (|\widetilde{\lambda}_{12}^3|^2 + |\widetilde{\lambda}_{22}^3|^2 + |\widetilde{\lambda}_{12}^2|^2) \end{aligned} \tag{2.99}$$

Note that $\delta_1 + \delta_2 = \varphi_1 + \varphi_2 + \varphi_3$, then by (2.96), it is easy to show that

$$\frac{2 \cdot \min\{\delta_1, \delta_2\}}{3} \leq \varphi_i \leq 2 \cdot \max\{\delta_1, \delta_2\} \tag{2.100}$$

which results in

$$
\begin{aligned}
P(\mathbf{d} &\to \overline{\mathbf{d}}|\widehat{\mathbf{H}}) \\
&\leq \exp\left(-\frac{\rho(\Phi(1,1)^2(|\widetilde{\lambda}^i_{11}|^2 + |\widetilde{\lambda}^i_{21}|^2) + \Phi(2,1)^2(|\widetilde{\lambda}^i_{12}|^2 + |\widetilde{\lambda}^i_{22}|^2))}{8}\right) \\
&\leq \exp\left(-\frac{\rho\varphi_i}{8}\right) \leq \exp\left(-\frac{\rho \cdot \min\{\delta_1, \delta_2\}}{12}\right)
\end{aligned}
\tag{2.101}
$$

and therefore

$$
\begin{aligned}
P(\mathbf{d} \to \overline{\mathbf{d}}) = E\big[P(\mathbf{d} \to \overline{\mathbf{d}}|\widehat{\mathbf{H}})\big] &\leq E\left[\exp\left(-\frac{\rho \cdot \delta_2}{12}\right)\right]\Pr\{\delta_1 > \delta_2\} \\
&+ E\left[\exp\left(-\frac{\rho \cdot \delta_1}{12}\right)\right]\Pr\{\delta_1 < \delta_2\}
\end{aligned}
\tag{2.102}
$$

Let $\mathbf{V}^1, \mathbf{V}^2, \mathbf{V}^3$ denote the unitary matrices derived from the singular value decomposition in (2.30) respectively for the three cases. Conditioned on $\mathbf{V}^1, \mathbf{V}^2, \mathbf{V}^3$, it can be shown that $\widetilde{\lambda}^1_{11}, \widetilde{\lambda}^1_{21}, \widetilde{\lambda}^2_{21}, \widetilde{\lambda}^1_{12}, \widetilde{\lambda}^1_{22}, \widetilde{\lambda}^2_{22}$ are i.i.d complex Gaussian random variables with mean 0 and variance 1. The same claim holds for $\widetilde{\lambda}^3_{11}, \widetilde{\lambda}^3_{21}, \widetilde{\lambda}^2_{11}, \widetilde{\lambda}^3_{12}, \widetilde{\lambda}^3_{22}, \widetilde{\lambda}^2_{12}$ as well. Then similar to (2.72), we have

$$
\begin{aligned}
E\left[\exp\left(-\frac{\rho \cdot \delta_i}{12}\right)\right] &= E_{\mathbf{V}^1,\mathbf{V}^2,\mathbf{V}^3}\left[E\left[\exp\left(-\frac{\rho \cdot \delta_i}{12}\right)\right]\Big|\mathbf{V}^1, \mathbf{V}^2, \mathbf{V}^3\right] \\
&\leq \frac{1}{\prod_{j=1}^{2}[1 + (\rho|\phi_j|^2/12)]^3}
\end{aligned}
\tag{2.103}
$$

Substituting (2.103) in (2.102), at high SNRs, we get

$$P(\mathbf{d} \to \overline{\mathbf{d}}) \leq \left(\frac{\rho}{12}\right)^{-6} \prod_{j=1}^{2} |\phi_j|^{-6} \tag{2.104}$$

As a result, the diversity $d \geq 6$. Similarly we can prove that $d \leq 6$. Therefore, $d = 6$ and we can achieve full diversity for User 1.

Now we prove that we can also achieve full diversity for User 2. Similar to (2.50), when there are 3 receive antennas, the channel equations can be written as

$$\begin{pmatrix} y_1^1 \\ (y_1^2)^* \\ y_2^1 \\ (y_2^2)^* \\ y_3^1 \\ (y_3^2)^* \end{pmatrix} = \sqrt{E_s} \begin{pmatrix} \widehat{h}_{11}^1 & \widehat{h}_{11}^1 & \widehat{g}_{11}^1 & \widehat{g}_{11}^1 \\ (\widehat{h}_{12}^2)^* & -(\widehat{h}_{12}^2)^* & (\widehat{g}_{12}^2)^* & -(\widehat{g}_{12}^2)^* \\ \widehat{h}_{21}^1 & \widehat{h}_{21}^1 & \widehat{g}_{21}^1 & \widehat{g}_{21}^1 \\ (\widehat{h}_{22}^2)^* & -(\widehat{h}_{22}^2)^* & (\widehat{g}_{22}^2)^* & -(\widehat{g}_{22}^2)^* \\ \widehat{h}_{31}^1 & \widehat{h}_{31}^1 & \widehat{g}_{31}^1 & \widehat{g}_{31}^1 \\ (\widehat{h}_{32}^2)^* & -(\widehat{h}_{32}^2)^* & (\widehat{g}_{32}^2)^* & -(\widehat{g}_{32}^2)^* \end{pmatrix} \begin{pmatrix} c_1 \\ c_2 \\ s_1 \\ s_2 \end{pmatrix} + \begin{pmatrix} n_1^1 \\ (n_1^2)^* \\ n_2^1 \\ (n_2^2)^* \\ n_3^1 \\ (n_3^2)^* \end{pmatrix}$$

$$(2.105)$$

By the method proposed above, we can detect the signals of User 1 with full diversity.
Here we let $\begin{pmatrix} \widehat{c}_1 \\ \widehat{c}_2 \end{pmatrix}$ denote the detected signals of User 1. We subtract the term of
$\begin{pmatrix} \widehat{h}_{11}^1 & \widehat{h}_{11}^1 \\ (\widehat{h}_{12}^2)^* & -(\widehat{h}_{12}^2)^* \\ \widehat{h}_{21}^1 & \widehat{h}_{21}^1 \\ (\widehat{h}_{22}^2)^* & -(\widehat{h}_{22}^2)^* \\ \widehat{h}_{31}^1 & \widehat{h}_{31}^1 \\ (\widehat{h}_{32}^2)^* & -(\widehat{h}_{32}^2)^* \end{pmatrix} \begin{pmatrix} \widehat{c}_1 \\ \widehat{c}_2 \end{pmatrix}$ from the channel equation to remove the effect of User
1 and will have

$$\begin{pmatrix} y_1^1 \\ (y_1^2)^* \\ y_2^1 \\ (y_2^2)^* \\ y_3^1 \\ (y_3^2)^* \end{pmatrix} - \begin{pmatrix} \widehat{h}_{11}^1 & \widehat{h}_{11}^1 \\ (\widehat{h}_{12}^2)^* & -(\widehat{h}_{12}^2)^* \\ \widehat{h}_{21}^1 & \widehat{h}_{21}^1 \\ (\widehat{h}_{22}^2)^* & -(\widehat{h}_{22}^2)^* \\ \widehat{h}_{31}^1 & \widehat{h}_{31}^1 \\ (\widehat{h}_{32}^2)^* & -(\widehat{h}_{32}^2)^* \end{pmatrix} \begin{pmatrix} \widehat{c}_1 \\ \widehat{c}_2 \end{pmatrix}$$

$$= \begin{pmatrix} \widehat{g}_{11}^1 & \widehat{g}_{11}^1 \\ (\widehat{g}_{12}^2)^* & -(\widehat{g}_{12}^2)^* \\ \widehat{g}_{21}^1 & \widehat{g}_{21}^1 \\ (\widehat{g}_{22}^2)^* & -(\widehat{g}_{22}^2)^* \\ \widehat{g}_{31}^1 & \widehat{g}_{31}^1 \\ (\widehat{g}_{32}^2)^* & -(\widehat{g}_{32}^2)^* \end{pmatrix} \begin{pmatrix} s_1 \\ s_2 \end{pmatrix} + \begin{pmatrix} n_{1R}^i \\ n_{2R}^i \\ n_{3R}^i \\ n_{1I}^i \\ n_{2I}^i \\ n_{3I}^i \end{pmatrix} + \sigma \qquad (2.106)$$

where $\sigma = \begin{pmatrix} \widehat{h}_{11}^1 & \widehat{h}_{11}^1 \\ (\widehat{h}_{12}^2)^* & -(\widehat{h}_{12}^2)^* \\ \widehat{h}_{21}^1 & \widehat{h}_{21}^1 \\ (\widehat{h}_{22}^2)^* & -(\widehat{h}_{22}^2)^* \\ \widehat{h}_{31}^1 & \widehat{h}_{31}^1 \\ (\widehat{h}_{32}^2)^* & -(\widehat{h}_{32}^2)^* \end{pmatrix} \left(\begin{pmatrix} c_1 \\ c_2 \end{pmatrix} - \begin{pmatrix} \widehat{c}_1 \\ \widehat{c}_2 \end{pmatrix} \right)$ denotes the residual error. Then

we can multiply both sides of Eq. (2.106) by $\begin{pmatrix} \widehat{g}_{11}^1 & \widehat{g}_{11}^1 \\ (\widehat{g}_{12}^2)^* & -(\widehat{g}_{12}^2)^* \\ \widehat{g}_{21}^1 & \widehat{g}_{21}^1 \\ (\widehat{g}_{22}^2)^* & -(\widehat{g}_{22}^2)^* \\ \widehat{g}_{31}^1 & \widehat{g}_{31}^1 \\ (\widehat{g}_{32}^2)^* & -(\widehat{g}_{32}^2)^* \end{pmatrix}^{\dagger}$ and use the same

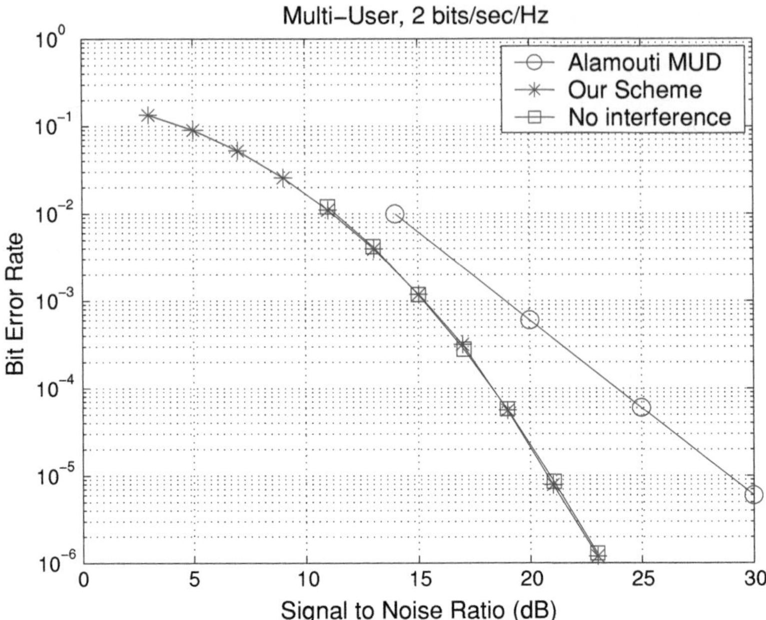

Fig. 2.4 Comparison of our scheme and Alamouti MUD for 2 users each with 2 transmit antennas

method in Sect. 2.2 to detect the signals of User 2. In what follows, we show that the method provides full diversity to User 2. There are two factors that result in error for User 2. The first one is the fading in the channel of User 2 and the second one is the error in detecting the symbols of User 1, i.e., error propagation. Let $\Pr(\mathbf{d}_2 \rightarrow \overline{\mathbf{d}_2})$ denote the pairwise error probability for User 2, we separate these two events to have

$$\Pr(\mathbf{d}_2 \rightarrow \overline{\mathbf{d}_2}) = \Pr\{\mathbf{d}_2 \rightarrow \overline{\mathbf{d}_2}|\sigma = 0\}\Pr\{\sigma = 0\} + \Pr\{\mathbf{d}_2 \rightarrow \overline{\mathbf{d}_2}|\sigma \neq 0\}\Pr\{\sigma \neq 0\}$$
$$= \Pr\{\mathbf{d}_2 \rightarrow \overline{\mathbf{d}_2}|\sigma = 0\}(1 - \Pr\{\sigma \neq 0\}) + \Pr\{\mathbf{d}_2 \rightarrow \overline{\mathbf{d}_2}|\sigma \neq 0\}\Pr\{\sigma \neq 0\} \quad (2.107)$$

Since $\Pr\{\mathbf{d}_2 \rightarrow \overline{\mathbf{d}_2}|\sigma \neq 0\} \leq 1$ and $1 - \Pr\{\sigma \neq 0\} \leq 1$, we have

$$\Pr(\mathbf{d}_2 \rightarrow \overline{\mathbf{d}_2}) \leq \Pr\{\mathbf{d}_2 \rightarrow \overline{\mathbf{d}_2}|\sigma = 0\}(1 - \Pr\{\sigma \neq 0\}) + \Pr\{\sigma \neq 0\}$$
$$\leq \Pr\{\mathbf{d}_2 \rightarrow \overline{\mathbf{d}_2}|\sigma = 0\} + \Pr\{\sigma \neq 0\} \quad (2.108)$$

Note that when $\sigma = 0$, we can follow the steps in Sect. 2.2 to detect the signals of User 2 and by the same technique used in Sect. 2.3, we can easily derive

$$\Pr\{\mathbf{d}_2 \rightarrow \overline{\mathbf{d}_2}|\sigma = 0\} \leq (\frac{\rho}{8})^{-6} \prod_{j=1}^{2} |\Phi(j, 1)|^{-6} = \tau_1 \rho^{-6} \quad (2.109)$$

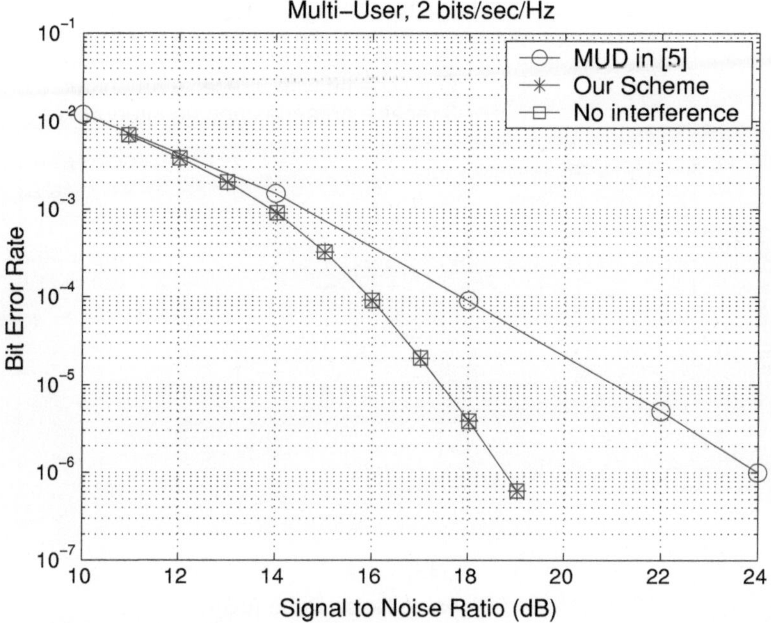

Fig. 2.5 Comparison of our scheme and MUD in [5] for 2 users each with 4 transmit antennas

where τ_1 is a constant. From (2.104), we know that

$$\Pr\{\sigma \neq 0\} \leq \tau_2 \rho^{-6} \tag{2.110}$$

where τ_2 is a constant. Substituting (2.109) and (2.110) in (2.108), we get

$$\Pr(\mathbf{d}_2 \rightarrow \bar{\mathbf{d}}_2) \leq (\tau_1 + \tau_1)\rho^{-6} \tag{2.111}$$

Using (2.111), it is easy to show that the diversity $d \geq 6$. Also we can show that diversity $d \leq 6$. So the diversity for User 2 is 6, i.e., full diversity. Therefore, we can achieve full diversity for both Users 1 and 2 which can also be confirmed by the simulations in the next section.

Note that when we complete the detection of the symbols of User 2, we can remove the effects of User 2 using the detected symbols of User 2 and re-detect the symbols of User 1. Simulation results show that such an iteration improves the coding gain. Finally, a similar antenna selection method at the receiver results in a diversity of NM for a general case of N transmit and M receive antennas.

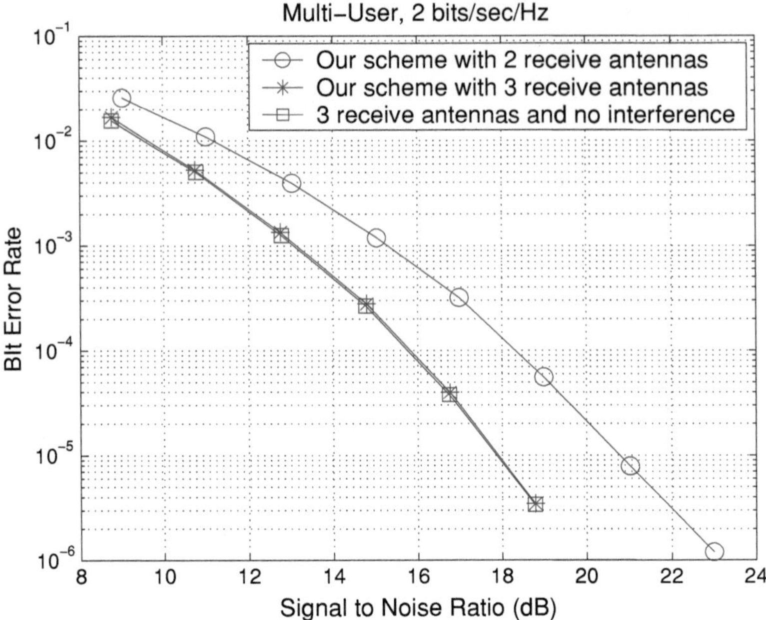

Fig. 2.6 Comparison of our scheme for 2 users each with 2 transmit antennas and different number of receive antennas

2.7 Simulation Results

In this section, we provide simulation results that confirm our analysis in the previous sections. We assume a quasi-static Rayleigh channel. The performance of our proposed scheme is shown in Figs. 2.4, 2.5 and 2.6. In each figure, the curves for Users 1 and 2 are identical. In Fig. 2.4, we consider 2 users each equipped with 2 transmit antennas and a receiver with 2 receive antennas. We compare our results using QPSK with the results in [5] for the same configuration without channel information at the transmitter. With 2 receive antennas, the multi-user detection (MUD) method offered in [5] can cancel the interference and provides a diversity of 2. Our scheme can also cancel the interference completely but provides a diversity of 4 by utilizing the channel information at the transmitter. We also present the results for a system with no interference. This is the same system when User 2 does not exist and can be easily achieved by $\mathbf{G} = 0$. Simulation results confirm that we have achieved interference cancellation completely.

Next, we present results for 2 users each with 4 transmit antennas and one receiver with 2 receive antennas in Fig. 2.5. We compare the performance of our method with the multiuser detection method in [5] using QOSTBC. As shown in Fig. 2.5, our scheme can achieve a diversity of 8, i.e., full diversity, by using channel information, while the MUD method using QOSTBC with no channel information can only achieve a diversity of 4.

Further, we show the results for 2 users each with 2 transmit antennas and one receiver with 2 or 3 receive antennas in Fig. 2.6. By increasing the number of receive antennas from 2 to 3, the diversity increases from 4 to 6. Therefore, extra receive antennas will provide extra diversity and the resulting diversity of the system is NM which confirms our theoretical analysis.

2.8 Conclusions

We have considered interference cancellation for a system with two users when users know each other channels. The goal is to utilize the channel information to cancel the interference without sacrificing the diversity or the complexity of the system. We have proposed a system to achieve the maximum possible diversity of NM with low complexity for 2 users each with N transmit antennas and one receiver with M receive antennas. To the best of our knowledge, this is the first multiuser detection scheme that achieves full diversity while providing a linear low complexity decoding. Our main idea is to design precoders, using the channel information, to make it possible for different users to transmit over orthogonal spaces. Then, using the orthogonality of the transmitted signals, the receiver can separate them and decode the signals independently. We have analytically proved that the system provides full diversity to both users. In addition, we provide simulation results that confirm our analytical results.

References

1. Ghaderipoor, A., Tellambura, C.: Optimal precoder for rate less than one space-time block codes. In: Proceedings of IEEE International Conference on Communication, Glasgow, Scotland (2007).
2. Simon, M.K., Alouini, M.-S.: Digital Communications over Fading Channels, 1st edn. Wiley, New York (2000)
3. Davis, P.J.: Circulant Matrices, 1st edn. Wiley, New York (1979)
4. Bayer-Fluckiger, E., Oggier, F., Viterbo, E.: New algebraic constructions of rotated Z^n-lattice constellations for the Rayleigh fading channel. IEEE Trans. Inf. Theory **50**, 702–714 (2004)
5. Kazemitabar, J., Jafarkhani, H.: Multiuser interference cancellation and detection for users with more than two transmit antennas. IEEE Trans. Commun. **56**(4), 574–583 (2008)

Chapter 3
Interference Cancellation and Detection for More than Two Users

3.1 Interference Cancellation for Four Users Each with Four Transmit Antennas

In this chapter, we assume a quasi-static flat Rayleigh fading channel model. The path gains are independent complex Gaussian random variables and are fixed during the transmission of one block. In addition, a short-term power constraint is assumed. For the sake of simplicity, we only present the scheme for four users each with four transmit antennas and one receiver with four receive antennas. By adjusting the dimensions of channel matrices, our proposed scheme can be easily applied to J users with J transmit antennas and one receiver with J receive antennas.

The block diagram of the system is shown in Fig. 3.1. We assume the channel matrices for Users 1, 2, 3, 4 are

$$
\begin{aligned}
\mathbf{H}_1 &= [h_1(i, j)]_{4\times4}, \\
\mathbf{H}_2 &= [h_2(i, j)]_{4\times4}, \\
\mathbf{H}_3 &= [h_3(i, j)]_{4\times4}, \\
\mathbf{H}_4 &= [h_4(i, j)]_{4\times4}
\end{aligned}
\tag{3.1}
$$

respectively. At the lth time slot, $l = 1, 2, 3, 4$, the precoders for Users 1, 2, 3, 4 are

$$
\begin{aligned}
\mathbf{A}_1^l &= [a_1^l(i, j)]_{4\times4}, \\
\mathbf{A}_2^l &= [a_2^l(i, j)]_{4\times4}, \\
\mathbf{A}_3^l &= [a_3^l(i, j)]_{4\times4}, \\
\mathbf{A}_4^l &= [a_4^i(i, j)]_{4\times4}
\end{aligned}
\tag{3.2}
$$

respectively. In every four time slots, Users 1, 2, 3, 4 send Quasi Orthogonal Space-Time Block Codes (QOSTBCs) [1]

F. Li, *Interference Cancellation Using Space-Time Processing and Precoding Design*, Signals and Communication Technology, DOI: 10.1007/978-3-642-30712-6_1, © Springer-Verlag Berlin Heidelberg 2013

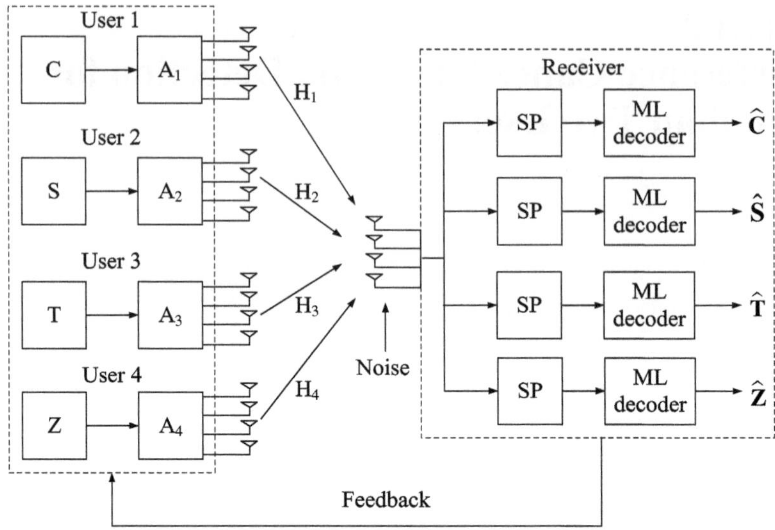

Fig. 3.1 Block diagram of the system

$$
\mathbf{C} = \begin{pmatrix} c_1 & -c_2^* & c_3 & -c_4^* \\ c_2 & c_1^* & c_4 & c_3^* \\ c_3 & -c_4^* & c_1 & -c_2^* \\ c_4 & c_3^* & c_2 & c_1^* \end{pmatrix}, \quad
\mathbf{S} = \begin{pmatrix} s_1 & -s_2^* & s_3 & -s_4^* \\ s_2 & s_1^* & s_4 & s_3^* \\ s_3 & -s_4^* & s_1 & -s_2^* \\ s_4 & s_3^* & s_2 & s_1^* \end{pmatrix}
$$

$$
\mathbf{T} = \begin{pmatrix} t_1 & -t_2^* & t_3 & -t_4^* \\ t_2 & t_1^* & t_4 & t_3^* \\ t_3 & -t_4^* & t_1 & -t_2^* \\ t_4 & t_3^* & t_2 & t_1^* \end{pmatrix}, \quad
\mathbf{Z} = \begin{pmatrix} z_1 & -z_2^* & z_3 & -z_4^* \\ z_2 & z_1^* & z_4 & z_3^* \\ z_3 & -z_4^* & z_1 & -z_2^* \\ z_4 & z_3^* & z_2 & z_1^* \end{pmatrix} \tag{3.3}
$$

respectively.

At time slot l, $l = 1, 2, 3, 4$, we have the following input-output equation

$$
\begin{aligned}
\mathbf{y}^l &= \sqrt{E_s}(\mathbf{H}_1\mathbf{A}_1^l\mathbf{c}(l) + \mathbf{H}_2\mathbf{A}_2^l\mathbf{s}(l) + \mathbf{H}_3\mathbf{A}_3^l\mathbf{t}(l) + \mathbf{H}_4\mathbf{A}_4^l\mathbf{z}(l)) + \mathbf{n}^l \\
&= \sqrt{E_s}(\mathbf{H}_1^l\mathbf{c}(l) + \mathbf{H}_2^l\mathbf{s}(l) + \mathbf{H}_3^l\mathbf{t}(l) + \mathbf{H}_4^l\mathbf{z}(l)) + \mathbf{n}^l
\end{aligned} \tag{3.4}
$$

where $\mathbf{H}_i^l = \mathbf{H}_i\mathbf{A}_i^l$ and $\mathbf{y}^l = \begin{pmatrix} y_1^l \\ y_2^l \\ y_3^l \\ y_4^l \end{pmatrix}$ denotes the received signals of the four receive

antennas at time slot l. E_s denotes the transmit energy of each user. $\mathbf{n}^l = \begin{pmatrix} n_1^l \\ n_2^l \\ n_3^l \\ n_4^l \end{pmatrix}$

denotes the noise at the receiver at time slot l. We assume that $n_1^l, n_2^l, n_3^l, n_4^l$ are i.i.d complex Gaussian noises with mean 0 and variance 1.

Applying some simple algebra to Equation (3.4), we have

$$
\mathbf{y}' = \sqrt{E_s}\left(\mathbf{H}'_1 \begin{pmatrix} c_1 \\ c_2 \\ c_3 \\ c_4 \end{pmatrix} + \mathbf{H}'_2 \begin{pmatrix} s_1 \\ s_2 \\ s_3 \\ s_4 \end{pmatrix} + \mathbf{H}'_3 \begin{pmatrix} t_1 \\ t_2 \\ t_3 \\ t_4 \end{pmatrix} + \mathbf{H}'_4 \begin{pmatrix} z_1 \\ z_2 \\ z_3 \\ z_4 \end{pmatrix} \right) + \mathbf{n}' \tag{3.5}
$$

where

$$
\mathbf{H}'_i = \begin{pmatrix}
h_i^1(1,1) & h_i^1(1,2) & h_i^1(1,3) & h_i^1(1,4) \\
h_i^1(2,1) & h_i^1(2,2) & h_i^1(2,3) & h_i^1(2,4) \\
h_i^1(3,1) & h_i^1(3,2) & h_i^1(3,3) & h_i^1(3,4) \\
h_i^1(4,1) & h_i^1(4,2) & h_i^1(4,3) & h_i^1(4,4) \\
(h_i^2(1,2))^* & -(h_i^2(1,1))^* & (h_i^2(1,4))^* & -(h_i^2(1,3))^* \\
(h_i^2(2,2))^* & -(h_i^2(2,1))^* & (h_i^2(2,4))^* & -(h_i^2(2,3))^* \\
(h_i^2(3,2))^* & -(h_i^2(3,1))^* & (h_i^2(3,4))^* & -(h_i^2(3,3))^* \\
(h_i^2(4,2))^* & -(h_i^2(4,1))^* & (h_i^2(4,4))^* & -(h_i^2(4,3))^* \\
h_i^3(1,3) & h_i^3(1,4) & h_i^3(1,1) & h_i^3(1,2) \\
h_i^3(2,3) & h_i^3(2,4) & h_i^3(2,1) & h_i^3(2,2) \\
h_i^3(3,3) & h_i^3(3,4) & h_i^3(3,1) & h_i^3(3,2) \\
h_i^3(4,3) & h_i^3(4,4) & h_i^3(4,1) & h_i^3(4,2) \\
(h_i^4(1,4))^* & -(h_i^4(1,3))^* & (h_i^4(1,2))^* & -(h_i^4(1,1))^* \\
(h_i^4(2,4))^* & -(h_i^4(2,3))^* & (h_i^4(2,2))^* & -(h_i^4(2,1))^* \\
(h_i^4(3,4))^* & -(h_i^4(3,3))^* & (h_i^4(3,2))^* & -(h_i^4(3,1))^* \\
(h_i^4(4,4))^* & -(h_i^4(4,3))^* & (h_i^4(4,2))^* & -(h_i^4(4,1))^*
\end{pmatrix},
$$

$$
\mathbf{y}' = \begin{pmatrix} \mathbf{y}^1 \\ (\mathbf{y}^2)^* \\ \mathbf{y}^3 \\ (\mathbf{y}^4)^* \end{pmatrix}, \quad
\mathbf{n}' = \begin{pmatrix} \mathbf{n}^1 \\ (\mathbf{n}^2)^* \\ \mathbf{n}^3 \\ (\mathbf{n}^4)^* \end{pmatrix} \tag{3.6}
$$

Now we choose precoders that can realize full diversity and interference cancellation for each user. First, we illustrate our main idea.

To realize interference cancellation, a straightforward idea is to transmit the symbols of the four users along four orthogonal directions. By doing so, it is easy to achieve interference cancellation at the receiver using zero-forcing. However, the difficulty lies in how to achieve full diversity as well. In [2], a scheme based on Alamouti structure has been proposed to achieve interference cancellation and full diversity for two users. When we have four users, the method does not work because four-dimensional rate-one complex orthogonal designs do not exist. An alternative is to use the quasi-orthogonal structure, but it cannot achieve full interference cancellation for each user due to its non-orthogonality.

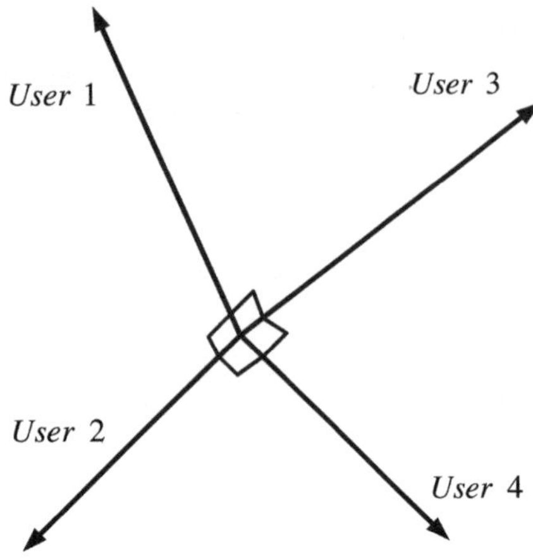

Fig. 3.2 Orthogonal structure of signal vectors in 4-dimensional space

To tackle all the above problems together, we propose a new precoder design scheme as follows. At each of the first 2 time slots, we design precoders such that symbols of User 1 and symbols of User 2 are transmitted along two orthogonal directions, respectively, as illustrated in Fig. 3.2. In addition, because of the characteristic of our designed precoders, each element of the equivalent channel matrices for Users 1 and 2 is still Gaussian. This property is critical to achieve full diversity for Users 1 and 2 as we will show later. Then we design precoders for Users 3 and 4, such that the transmit directions of their signals are orthogonal to each other. Note that it is impossible to obtain this orthogonal structure and make each element of the equivalent channel matrices for Users 3 and 4 still Gaussian. This is the main difference between the precoders for Users 1, 2 and the precoders for Users 3, 4, at the first 2 time slots.

At the second 2 time slots, we also design precoders to make the transmit directions of signals orthogonal to each other. However, we design the precoders for Users 3 and 4 first, such that each element of the equivalent channel matrices for Users 3 and 4 is Gaussian. Then we design the precoders for Users 1 and 2 to obtain the orthogonal structure. As a result, elements of the equivalent channel matrices for Users 1 and 2 will not be Gaussian at the second 2 time slots. Later we will prove that by using such precoders, we can achieve interference cancellation and full diversity for each user. In what follows, we will describe the details of our precoder designs.

At time slot 1, in order to have orthogonality between User 1 and User 2, we design the precoders such that

$$
\begin{pmatrix} h_2^1(1,1) \\ h_2^1(2,1) \\ h_2^1(3,1) \\ h_2^1(4,1) \end{pmatrix} = \eta \begin{pmatrix} -h_1^1(2,1) \\ h_1^1(1,1) \\ -h_1^1(4,1) \\ h_1^1(3,1) \end{pmatrix}^*
\tag{3.7}
$$

where $h_1^1(i, j)$ and $h_2^1(i, j)$ are elements of the equivalent channel matrices in Equation (3.6). Equation (3.7) can be rewritten as

$$\mathbf{H}_2 \begin{pmatrix} a_2^1(1, 1) \\ a_2^1(2, 1) \\ a_2^1(3, 1) \\ a_2^1(4, 1) \end{pmatrix} = \widehat{\mathbf{H}}_1^* \begin{pmatrix} a_1^1(1, 1) \\ a_1^1(2, 1) \\ a_1^1(3, 1) \\ a_1^1(4, 1) \end{pmatrix}^* \tag{3.8}$$

where

$$\widehat{\mathbf{H}}_1 = \begin{pmatrix} -h_1(2, 1) & -h_1(2, 2) & -h_1(2, 3) & -h_1(2, 4) \\ h_1(1, 1) & h_1(1, 2) & h_1(1, 3) & h_1(1, 4) \\ -h_1(4, 1) & -h_1(4, 2) & -h_1(4, 3) & -h_1(4, 4) \\ h_1(3, 1) & h_1(3, 2) & h_1(3, 3) & h_1(3, 4) \end{pmatrix} \tag{3.9}$$

Now let

$$\mathbf{Q} = \mathbf{H}_2^{-1} \widehat{\mathbf{H}}_1^* = \mathbf{U}\mathbf{\Sigma}\mathbf{V}^H \tag{3.10}$$

where we have made the singular value decomposition. It has been proved in [2] that

$$\begin{pmatrix} a_1^1(1, 1) \\ a_1^1(2, 1) \\ a_1^1(3, 1) \\ a_1^1(4, 1) \end{pmatrix} = \mathbf{v}(i)^*, \quad \begin{pmatrix} a_2^1(1, 1) \\ a_2^1(2, 1) \\ a_2^1(3, 1) \\ a_2^1(4, 1) \end{pmatrix} = \mathbf{u}(i), \quad \eta = \frac{1}{\Sigma(i, i)}, \quad i = 1, 2, 3, 4$$

$$\tag{3.11}$$

will satisfy Equation (3.8). There are four different choices for $\begin{pmatrix} a_1^1(1, 1) \\ a_1^1(2, 1) \\ a_1^1(3, 1) \\ a_1^1(4, 1) \end{pmatrix}$ and

$\begin{pmatrix} a_2^1(1, 1) \\ a_2^1(2, 1) \\ a_2^1(3, 1) \\ a_2^1(4, 1) \end{pmatrix}$ depending on which i we pick. Different choices of i result in different

performances. For given channel matrices $\mathbf{H}_1$ and $\mathbf{H}_2$, at time slot 1, we let $\mathbf{v}' = \mathbf{v}(i)^*$, $i \in \{1, 2, 3, 4\}$, such that the norm of $\mathbf{H}_1\mathbf{v}'$ is the largest, i.e.,

$$\mathbf{v}' = \arg \max_{\mathbf{v}(i)^*, i=1,2,3,4} \|\mathbf{H}_1\mathbf{v}(i)^*\|_F^2 \tag{3.12}$$

Then for User 1, at time slot 1, we let

$$\begin{pmatrix} a_1^1(1, 1) \\ a_1^1(2, 1) \\ a_1^1(3, 1) \\ a_1^1(4, 1) \end{pmatrix} = \frac{\mathbf{v}'}{\sqrt{1 + \sum_{j=1}^3 k_j^2}}, \quad \begin{pmatrix} a_1^1(1, i') \\ a_1^1(2, i') \\ a_1^1(3, i') \\ a_1^1(4, i') \end{pmatrix} = k_{i'-1} \cdot \begin{pmatrix} a_1^1(1, 1) \\ a_1^1(2, 1) \\ a_1^1(3, 1) \\ a_1^1(4, 1) \end{pmatrix}, \quad i' = 2, 3, 4 \tag{3.13}$$

For User 2, at time slot 1, we let

$$\begin{pmatrix} a_2^1(1,1) \\ a_2^1(2,1) \\ a_2^1(3,1) \\ a_2^1(4,1) \end{pmatrix} = \frac{\mathbf{u}(i)}{\sqrt{1 + \sum_{j=1}^{3} k_j^2}},$$

$$\begin{pmatrix} a_2^1(1,i') \\ a_2^1(2,i') \\ a_2^1(3,i') \\ a_2^1(4,i') \end{pmatrix} = k_{i'-1} \begin{pmatrix} a_2^1(1,1) \\ a_2^1(2,1) \\ a_2^1(3,1) \\ a_2^1(4,1) \end{pmatrix}, \quad i' = 2,3,4 \tag{3.14}$$

where i is the same as that in Equation (3.12). As we will discuss later, we choose parameters k_1, k_2, k_3 to maximize the coding gain. The choice of k_1, k_2, k_3 will complete the precoder design for Users 1 and 2 at time slot 1. Note that the designed precoders $\mathbf{A}_1^1$, $\mathbf{A}_2^1$ satisfy $||\mathbf{A}_1^1||_F^2 = ||\mathbf{A}_2^1||_F^2 = 1$ and the signals of User 1 and User 2 will be transmitted along two orthogonal directions as shown in Fig. 3.2.

In order to derive the orthogonality among Users 1, 2, 3 at time slot 1, we design precoder $\mathbf{A}_3^1$ to satisfy the following properties:

1. Complex vectors $\mathbf{H}_1 \begin{pmatrix} a_1^1(1,1) \\ a_1^1(2,1) \\ a_1^1(3,1) \\ a_1^1(4,1) \end{pmatrix}$, $\mathbf{H}_2 \begin{pmatrix} a_2^1(1,1) \\ a_2^1(2,1) \\ a_2^1(3,1) \\ a_2^1(4,1) \end{pmatrix}$, $\mathbf{H}_3 \begin{pmatrix} a_3^1(1,1) \\ a_3^1(2,1) \\ a_3^1(3,1) \\ a_3^1(4,1) \end{pmatrix}$ are

orthogonal to each other.

2.

$$\begin{pmatrix} a_3^1(1,i) \\ a_3^1(2,i) \\ a_3^1(3,i) \\ a_3^1(4,i) \end{pmatrix} = k_{i-1} \begin{pmatrix} a_3^1(1,1) \\ a_3^1(2,1) \\ a_3^1(3,1) \\ a_3^1(4,1) \end{pmatrix}, \quad i = 2,3,4 \tag{3.15}$$

3. The Frobenius norm of complex matrix $\mathbf{A}_3^1$ is equal to 1.

In order to maximize the coding gain, $\mathbf{A}_3^1$ can be further chosen numerically such that the norm of $\mathbf{H}_3\mathbf{A}_3^1$ is maximized. Similarly, for User 4, at time slot 1, in order to derive the orthogonality as shown in Fig. 3.2, we choose precoder $\mathbf{A}_4^1$ to satisfy the following properties:

1. Complex vectors $\mathbf{H}_1 \begin{pmatrix} a_1^1(1,1) \\ a_1^1(2,1) \\ a_1^1(3,1) \\ a_1^1(4,1) \end{pmatrix}$, $\mathbf{H}_2 \begin{pmatrix} a_2^1(1,1) \\ a_2^1(2,1) \\ a_2^1(3,1) \\ a_2^1(4,1) \end{pmatrix}$, $\mathbf{H}_3 \begin{pmatrix} a_3^1(1,1) \\ a_3^1(2,1) \\ a_3^1(3,1) \\ a_3^1(4,1) \end{pmatrix}$,

$\mathbf{H}_4 \begin{pmatrix} a_4^1(1,1) \\ a_4^1(2,1) \\ a_4^1(3,1) \\ a_4^1(4,1) \end{pmatrix}$ are orthogonal to each other.

2.

$$
\begin{pmatrix} a_4^1(1, i) \\ a_4^1(2, i) \\ a_4^1(3, i) \\ a_4^1(4, i) \end{pmatrix} = k_{i-1} \cdot \begin{pmatrix} a_4^1(1, 1) \\ a_4^1(2, 1) \\ a_4^1(3, 1) \\ a_4^1(4, 1) \end{pmatrix}, \quad i = 2, 3, 4 \tag{3.16}
$$

3. The Frobenius norm of complex matrix $\mathbf{A}_4^1$ is equal to 1.

Similarly, in order to improve the coding gain, $\mathbf{A}_4^1$ can be further chosen numerically such that the norm of $\mathbf{H}_4 \mathbf{A}_4^1$ is maximized. By choosing $\mathbf{A}_1^1$, $\mathbf{A}_2^1$, $\mathbf{A}_3^1$, $\mathbf{A}_4^1$, the precoder design at time slot 1 is complete.

At time slot 2, the precoder design is similar to that at time slot 1. The difference is that we choose $\mathbf{u}' = \mathbf{u}(i)$, $i \in \{1, 2, 3, 4\}$, such that $||\mathbf{H}_2 \mathbf{u}'||_F$ is the largest, i.e.,

$$
\mathbf{u}' = \arg \max_{\mathbf{u}(i), i=1,2,3,4} ||\mathbf{H}_2 \mathbf{u}(i)||_F^2 \tag{3.17}
$$

Then we let

$$
\begin{pmatrix} a_2^2(1, 1) \\ a_2^2(2, 1) \\ a_2^2(3, 1) \\ a_2^2(4, 1) \end{pmatrix} = \frac{\mathbf{u}'}{\sqrt{1 + \sum_{j=1}^3 k_j^2}}, \quad \begin{pmatrix} a_2^2(1, i') \\ a_2^2(2, i') \\ a_2^2(3, i') \\ a_2^2(4, i') \end{pmatrix} = k_{i'-1} \begin{pmatrix} a_2^2(1, 1) \\ a_2^2(2, 1) \\ a_2^2(3, 1) \\ a_2^2(4, 1) \end{pmatrix}, \quad i' = 2, 3, 4 \tag{3.18}
$$

For User 1, at time slot 2, we choose

$$
\begin{pmatrix} a_1^2(1, 1) \\ a_1^2(2, 1) \\ a_1^2(3, 1) \\ a_1^2(4, 1) \end{pmatrix} = \frac{\mathbf{v}(i)}{\sqrt{1 + \sum_{j=1}^3 k_j^2}}, \quad \begin{pmatrix} a_1^2(1, i') \\ a_1^2(2, i') \\ a_1^2(3, i') \\ a_1^2(4, i') \end{pmatrix} = k_{i'-1} \begin{pmatrix} a_1^2(1, 1) \\ a_1^2(2, 1) \\ a_1^2(3, 1) \\ a_1^2(4, 1) \end{pmatrix}, \quad i' = 2, 3, 4 \tag{3.19}
$$

where i is the same with that in Equation (3.17).

Design of $\mathbf{A}_3^2$, $\mathbf{A}_4^2$ is similar to that of $\mathbf{A}_3^1$, $\mathbf{A}_4^1$. By switching the terms related to Users 1 and 2 with those of Users 3 and 4, respectively, we can design the precoders at time slots 3 and 4.

Till now, the precoder design for each user at the first 4 time slots is complete. When there are J users, at time slots $2k - 1$ and $2k$, we first design precoders for Users $2k - 1$ and $2k$ similar to what we do for Users 1 and 2. Then we design precoders for other users such that all of them transmit along orthogonal directions. Therefore, the above idea for 4 users can be easily extended to any number of users. In the next two sections, we will illustrate how to decode and why our scheme can realize interference cancellation and full diversity for each user.

3.2 Decoding

Using our precoders, Equation (3.5) becomes

$$\overline{\mathbf{y}} = \sqrt{E_s}(\overline{\mathbf{H}}_1 \begin{pmatrix} c_1 \\ c_2 \\ c_3 \\ c_4 \end{pmatrix} + \overline{\mathbf{H}}_2 \begin{pmatrix} s_1 \\ s_2 \\ s_3 \\ s_4 \end{pmatrix} + \overline{\mathbf{H}}_3 \begin{pmatrix} t_1 \\ t_2 \\ t_3 \\ t_4 \end{pmatrix} + \overline{\mathbf{H}}_4 \begin{pmatrix} z_1 \\ z_2 \\ z_3 \\ z_4 \end{pmatrix}) + \overline{\mathbf{n}} \qquad (3.20)$$

where

$$\overline{\mathbf{H}}_i = \begin{pmatrix}
h_i^1(1,1) & k_1 h_i^1(1,1) & k_2 h_i^1(1,1) & k_3 h_i^1(1,1) \\
h_i^1(2,1) & k_1 h_i^1(2,1) & k_2 h_i^1(2,1) & k_3 h_i^1(2,1) \\
h_i^1(3,1) & k_1 h_i^1(3,1) & k_2 h_i^1(3,1) & k_3 h_i^1(3,1) \\
h_i^1(4,1) & k_1 h_i^1(4,1) & k_2 h_i^1(4,1) & k_3 h_i^1(4,1) \\
k_1(h_i^2(1,1))^* & -(h_i^2(1,1))^* & k_3(h_i^2(1,1))^* & -k_2(h_i^2(1,1))^* \\
k_1(h_i^2(2,1))^* & -(h_i^2(2,1))^* & k_3(h_i^2(2,1))^* & -k_2(h_i^2(2,1))^* \\
k_1(h_i^2(3,1))^* & -(h_i^2(3,1))^* & k_3(h_i^2(3,1))^* & -k_2(h_i^2(3,1))^* \\
k_1(h_i^2(4,1))^* & -(h_i^2(4,1))^* & k_3(h_i^2(4,1))^* & -k_2(h_i^2(4,1))^* \\
k_2 h_i^3(1,1) & k_3 h_i^3(1,1) & h_i^3(1,1) & k_1 h_i^3(1,1) \\
k_2 h_i^3(2,1) & k_3 h_i^3(2,1) & h_i^3(2,1) & k_1 h_i^3(2,1) \\
k_2 h_i^3(3,1) & k_3 h_i^3(3,1) & h_i^3(3,1) & k_1 h_i^3(3,1) \\
k_2 h_i^3(4,1) & k_3 h_i^3(4,1) & h_i^3(4,1) & k_1 h_i^3(4,1) \\
k_3(h_i^4(1,1))^* & -k_2(h_i^4(1,1))^* & k_1(h_i^4(1,1))^* & -(h_i^4(1,1))^* \\
k_3(h_i^4(2,1))^* & -k_2(h_i^4(2,1))^* & k_1(h_i^4(2,1))^* & -(h_i^4(2,1))^* \\
k_3(h_i^4(3,1))^* & -k_2(h_i^4(3,1))^* & k_1(h_i^4(3,1))^* & -(h_i^4(3,1))^* \\
k_3(h_i^4(4,1))^* & -k_2(h_i^4(4,1))^* & k_1(h_i^4(4,1))^* & -(h_i^4(4,1))^*
\end{pmatrix} \qquad (3.21)$$

Here $\overline{\mathbf{y}}$ and $\overline{\mathbf{n}}$ are the same with $\mathbf{y}'$ and $\mathbf{n}'$ in Equation (3.5). Note that using our precoders, each column of array $\overline{\mathbf{H}}_1$ is orthogonal to each column of matrices $\overline{\mathbf{H}}_2$, $\overline{\mathbf{H}}_3$, $\overline{\mathbf{H}}_4$.

In order to decode symbols from User 1, we multiply both sides of Equation (3.20) by array $\overline{\mathbf{H}}_1^\dagger$ to achieve

$$\overline{\mathbf{H}}_1^\dagger \overline{\mathbf{y}} = \sqrt{E_s}\overline{\mathbf{H}}_1^\dagger \overline{\mathbf{H}}_1 \begin{pmatrix} c_1 \\ c_2 \\ c_3 \\ c_4 \end{pmatrix} + \overline{\mathbf{H}}_1^\dagger \overline{\mathbf{n}} \qquad (3.22)$$

Note that the noise elements of $\overline{\mathbf{H}}_1^\dagger \overline{\mathbf{n}}$ are correlated with covariance matrix $\overline{\mathbf{H}}_1^\dagger \overline{\mathbf{H}}_1$. We can whiten this noise vector by multiplying both sides of Equation (3.22) by the matrix $(\overline{\mathbf{H}}_1^\dagger \overline{\mathbf{H}}_1)^{-\frac{1}{2}}$ as follows

$$(\overline{\mathbf{H}}_1^\dagger \overline{\mathbf{H}}_1)^{-\frac{1}{2}} \overline{\mathbf{H}}_1^\dagger \overline{\mathbf{y}} = \sqrt{E_s}(\overline{\mathbf{H}}_1^\dagger \overline{\mathbf{\Pi}}_1)^{\frac{1}{2}} \begin{pmatrix} c_1 \\ c_2 \\ c_3 \\ c_4 \end{pmatrix} + \widehat{\mathbf{n}} \tag{3.23}$$

where $\widehat{\mathbf{n}} = (\overline{\mathbf{H}}_1^\dagger \overline{\mathbf{H}}_1)^{-\frac{1}{2}}(\overline{\mathbf{H}}_1^\dagger \overline{\mathbf{n}})$ has uncorrelated elements $\sim CN(0, 1)$. Equation (3.23) can be further rewritten as

$$(\overline{\mathbf{H}}_1^\dagger \overline{\mathbf{H}}_1)^{-\frac{1}{2}} \overline{\mathbf{H}}_1^\dagger \overline{\mathbf{y}} = \sqrt{E_s} \begin{pmatrix} x_1 & x_2 & x_3 & x_4 \\ x_2 & x_5 & x_6 & x_7 \\ x_3 & x_6 & x_8 & x_9 \\ x_4 & x_7 & x_9 & x_{10} \end{pmatrix}^{\frac{1}{2}} \begin{pmatrix} c_1 \\ c_2 \\ c_3 \\ c_4 \end{pmatrix} + \widehat{\mathbf{n}} \tag{3.24}$$

where

$$\begin{aligned}
x_1 &= a + k_1^2 b + k_2^2 c + k_3^2 d, & x_2 &= k_1 a - k_1 b + k_2 k_3 c - k_2 k_3 d \\
x_3 &= k_2 a + k_1 k_3 b + k_2 c + k_1 k_3 d, & x_4 &= k_3 a - k_1 k_2 b + k_1 k_2 c - k_3 d \\
x_5 &= k_1^2 a + b + k_3^2 c + k_2^2 d, & x_6 &= k_1 k_2 a - k_3 b + k_3 c - k_1 k_2 d \\
x_7 &= k_1 k_3 a + k_2 b + k_1 k_3 c + k_2 d, & x_8 &= k_2^2 a + k_3^2 b + c + k_1^2 d \\
x_9 &= k_2 k_3 a - k_2 k_3 b + k_1 c - k_1 d, & x_{10} &= k_3^2 a + k_2^2 b + k_1^2 c + d
\end{aligned} \tag{3.25}$$

$$a = \sum_{i=1}^{4} |h_1^1(i, 1)|^2, \quad b = \sum_{i=1}^{4} |h_1^2(i, 1)|^2,$$

$$c = \sum_{i=1}^{4} |h_1^3(i, 1)|^2, \quad d = \sum_{i=1}^{4} |h_1^4(i, 1)|^2 \tag{3.26}$$

Now let

$$\widehat{\mathbf{H}} = \begin{pmatrix} x_1 & x_2 & x_3 & x_4 \\ x_2 & x_5 & x_6 & x_7 \\ x_3 & x_6 & x_8 & x_9 \\ x_4 & x_7 & x_9 & x_{10} \end{pmatrix}^{\frac{1}{2}} \tag{3.27}$$

From Equation (3.24), we can see that User 1 transmits 4 different codewords along 4 different equivalent channel vectors in the 4 time slots. So the rate is 1. If k_1, k_2, k_3 are all real, from (3.27), it is easy to see that the equivalent channel matrix $\widehat{\mathbf{H}}$ is real. So if QAM is used, Equation (3.24) is equivalent to the following two equations

$$\widehat{\mathbf{H}}^{-1} \mathrm{Real}\{\overline{\mathbf{H}}_1^\dagger \overline{\mathbf{y}}\} = \sqrt{E_s} \widehat{\mathbf{H}} \begin{pmatrix} c_{1R} \\ c_{2R} \\ c_{3R} \\ c_{4R} \end{pmatrix} + \mathrm{Real}\{\widehat{\mathbf{n}}\} \tag{3.28}$$

$$\widehat{\mathbf{H}}^{-1}\mathrm{Imag}\{\overline{\mathbf{H}}_1^{\dagger}\overline{\mathbf{y}}\} = \sqrt{E_s}\,\widehat{\mathbf{H}}\begin{pmatrix} c_{1I} \\ c_{2I} \\ c_{3I} \\ c_{4I} \end{pmatrix} + \mathrm{Imag}\{\widehat{\mathbf{n}}\} \tag{3.29}$$

Then we can use the Maximum-Likelihood method to detect the real and imaginary parts of these 4 codewords separately. For example, by Equation (3.28), we can detect $c_{1R}, \ldots, c_{4R}$ by

$$\begin{pmatrix} \widehat{c}_{1R} \\ \widehat{c}_{2R} \\ \widehat{c}_{3R} \\ \widehat{c}_{4R} \end{pmatrix} = \arg\min_{c_{1R},\ldots,c_{4R}} ||\widehat{\mathbf{H}}^{-1}\mathrm{Real}\{\overline{\mathbf{H}}_1^{\dagger}\overline{\mathbf{y}}\} - \sqrt{E_s}\,\widehat{\mathbf{H}}\begin{pmatrix} c_{1R} \\ c_{2R} \\ c_{3R} \\ c_{4R} \end{pmatrix}||_F^2 \tag{3.30}$$

Similarly, using Equation (3.29), we can detect $\begin{pmatrix} c_{1I} \\ c_{2I} \\ c_{3I} \\ c_{4I} \end{pmatrix}$. Note that the decoding complexity is pair-wise decoding. In order to detect codewords of Users 2, 3, 4, we can multiply both sides of Equation (3.20) with matrix $\overline{\mathbf{H}}_2^{\dagger}$, $\overline{\mathbf{H}}_3^{\dagger}$, $\overline{\mathbf{H}}_4^{\dagger}$, respectively, to remove the signals of other users and use a similar method to complete the decoding.

3.3 Proof of Full Diversity

In this section, we prove that we can achieve diversity 16, i.e., full diversity, using our proposed precoding scheme. We only present the proof for User 1, since the proof for Users 2, 3, 4 is the same. Diversity is defined as

$$d = -\lim_{\rho\to\infty}\frac{\log P_e}{\log \rho} \tag{3.31}$$

where ρ denotes the SNR and P_e represents the probability of error. We first consider (3.24) to analyze the diversity for User 1. Here we add a unitary rotation $\mathbf{R}$ to $\begin{pmatrix} c_1 \\ c_2 \\ c_3 \\ c_4 \end{pmatrix}$. Thus, the data vector is $\mathbf{d} = \mathbf{R}\begin{pmatrix} c_1 \\ c_2 \\ c_3 \\ c_4 \end{pmatrix}$ and we define the error vector $\boldsymbol{\varepsilon} = \begin{pmatrix} c_1 \\ c_2 \\ c_3 \\ c_4 \end{pmatrix} - \begin{pmatrix} \widehat{c}_1 \\ \widehat{c}_2 \\ \widehat{c}_3 \\ \widehat{c}_4 \end{pmatrix}$. By (3.24), the pairwise error probability (PEP) can be derived using the Gaussian tail function as [3]

$$P(\mathbf{d} \to \overline{\mathbf{d}}|\widehat{\mathbf{H}}) = Q\left(\sqrt{\frac{\rho\|\widehat{\mathbf{H}}\mathbf{R}\boldsymbol{\varepsilon}\|_F^2}{4}}\right)$$

$$= Q\left(\sqrt{\frac{\rho\boldsymbol{\varepsilon}^\dagger\mathbf{R}^\dagger\|\widehat{\mathbf{H}}\|^2\mathbf{R}\boldsymbol{\varepsilon}}{4}}\right)$$

$$\leq \exp\left(-\frac{\rho\boldsymbol{\varepsilon}^\dagger\mathbf{R}^\dagger\|\widehat{\mathbf{H}}\|^2\mathbf{R}\boldsymbol{\varepsilon}}{4}\right) \tag{3.32}$$

Now we assume $\mathbf{R}\boldsymbol{\varepsilon} = \begin{pmatrix} \gamma_1 \\ \gamma_2 \\ \gamma_3 \\ \gamma_4 \end{pmatrix}$. Substituting $\mathbf{R}\boldsymbol{\varepsilon}$ and $\widehat{\mathbf{H}}$ in Equation (3.27) into (3.32), we have

$$P(\mathbf{d} \to \overline{\mathbf{d}}|\widehat{\mathbf{H}}) \leq \exp\left(-\frac{\rho\zeta}{4}\right) \tag{3.33}$$

where

$$\zeta = a|\gamma_1 + k_1\gamma_2 + k_2\gamma_3 + k_3\gamma_4|^2 + b|k_3\gamma_1 - \gamma_2 + k_1\gamma_3 - k_2\gamma_4|^2 +$$
$$c|k_2\gamma_1 + k_3\gamma_2 + \gamma_3 + k_1\gamma_4|^2 + d|k_1\gamma_1 - k_2\gamma_2 + k_3\gamma_3 - \gamma_4|^2 \tag{3.34}$$

Further, we have

$$P(\mathbf{d} \to \overline{\mathbf{d}}|\widehat{\mathbf{H}}) \leq \exp\left(-\frac{\rho \cdot a|\gamma_1 + k_1\gamma_2 + k_2\gamma_3 + k_3\gamma_4|^2}{4}\right)$$

$$= \exp\left(-\frac{\rho \cdot \sum_{i=1}^{4}|h_1^1(i,1)|^2|\gamma_1 + k_1\gamma_2 + k_2\gamma_3 + k_3\gamma_4|^2}{4}\right) \tag{3.35}$$

Note that

$$\sum_{i=1}^{4}|h_1^1(i,1)|^2 = \|\mathbf{H}_1\mathbf{v}'\|^2 \geq \frac{1}{4}\|\mathbf{H}_1\mathbf{V}\|^2 = \frac{1}{4}\|\mathbf{H}_1\|^2 = \frac{1}{4}\sum_{i=1}^{4}\sum_{j=1}^{4}|h_1(i,j)|^2 \tag{3.36}$$

So we have

$$P(\mathbf{d} \to \overline{\mathbf{d}}) = E[P(\mathbf{d} \to \overline{\mathbf{d}}|\widehat{\mathbf{H}})] \leq E\left[\exp\left(-\frac{\rho \cdot \sum_{i=1}^{4}\sum_{j=1}^{4}|h_1(i,j)|^2 \cdot \zeta'}{16}\right)\right]$$

$$= \frac{1}{\prod_{j=1}^{16}[1 + (\rho\zeta'/16)]} \tag{3.37}$$

where

$$\zeta' = \frac{|\gamma_1 + k_1\gamma_2 + k_2\gamma_3 + k_3\gamma_4|^2}{1 + \sum_{j=1}^{3} k_j^2} \tag{3.38}$$

At high SNRs, one can neglect the one in the denominator and get

$$P(\mathbf{d} \to \overline{\mathbf{d}}) \leq \left(\frac{\rho\zeta'}{16}\right)^{-16} \tag{3.39}$$

Then, it is easy to show that the diversity is 16 if we simply choose $\mathbf{KR}$ such that

$$\zeta' = \mathbf{KR}\varepsilon \neq 0 \tag{3.40}$$

where $\mathbf{K} = \frac{1}{\sqrt{1+\sum_{j=1}^{3} k_j^2}} \begin{pmatrix} 1 & k_1 & k_2 & k_3 \end{pmatrix}$ is a normalized vector. Therefore, by using our scheme, User 1 can achieve full diversity. In addition, in order to maximize the coding gain, we need to choose $\mathbf{KR}$ such that the minimum possible norm of $\mathbf{KR}\varepsilon$ is maximized. For QAM, it is not hard to do so. For example, when QPSK is adopted, we can simply choose $\mathbf{KR} = \frac{1}{85} \begin{pmatrix} 1 & 2 & 4 & 8 \end{pmatrix}$. It is easy to check that the minimum possible norm of $\mathbf{KR}\varepsilon$ is maximized. Similarly, we can also prove that the diversity for Users 2, 3, 4, is 16 as well. Therefore, our scheme can achieve full diversity for each user. When we use Equations (3.28), (3.29) to simplify the decoding complexity, similar techniques can be used to complete the proof of full diversity. Note that our precoding design procedure itself does not rely on the channel statistics. So using our scheme, the pairwise error probability can always be upper bounded by

$$P(\mathbf{d} \to \overline{\mathbf{d}}) \leq E\left[\exp\left(-\frac{\rho \cdot \|\mathbf{H}_1\|_F^2 \cdot \zeta'}{16}\right)\right] \tag{3.41}$$

like Equation (3.37), where $\mathbf{H}_1$ is the channel matrix for User 1. This means that the proposed procedure is universal in that it can achieve the maximum possible diversity over any fading distribution.

3.4 Extension to J Users with N Transmit Antennas and One Receiver with M Receive Antennas

In this section, we show that the presented scheme can be extended to a general case of J users each with N transmit antennas and one receiver with M receive antennas. For the simplification of presentation, we discuss 3 cases where among parameters M, N and J, two are the same and the third one is larger than the other two. It is easy to extend the results to a general case. In addition, we just show our schemes when

J, M and N are all even. By some simple antenna or user removals, our proposed scheme can also be used when not all of J, M and N are even.

3.4.1 More Transmit Antennas, i.e., $N > J = M$

First, we consider the case $N > J = M$. For simplification, we assume $N = 6$, $J = M = 4$. Also we take User 1 for example. Each user transmits QOSTBCs and the precoder will be a 6×4 matrix. Similar to Equation (3.7), to make User 1 and User 2 orthogonal to each other, we design precoders such that

$$
\mathbf{H}_2
\begin{pmatrix}
a_2^1(1,1) \\
a_2^1(2,1) \\
a_2^1(3,1) \\
a_2^1(4,1) \\
a_2^1(5,1) \\
a_2^1(6,1)
\end{pmatrix}
= \eta \widehat{\mathbf{H}}_1^*
\begin{pmatrix}
a_1^1(1,1) \\
a_1^1(2,1) \\
a_1^1(3,1) \\
a_1^1(4,1) \\
a_1^1(5,1) \\
a_1^1(6,1)
\end{pmatrix}^{*}
\tag{3.42}
$$

where $\widehat{\mathbf{H}}_1$ and $\mathbf{H}_2$ are all 4×6 matrices. Since $\widehat{\mathbf{H}}_1$ and $\mathbf{H}_2$ are not square matrices, we cannot take the inverse as in Equation (3.10). Instead, we multiply both sides of Equation (3.42) by $\mathbf{H}_2^\dagger$ and $(\mathbf{H}_2^\dagger \mathbf{H}_2)^{-1}$ resulting in

$$
\mathbf{Q} = (\mathbf{H}_2^\dagger \mathbf{H}_2)^{-1} \mathbf{H}_2^\dagger \widehat{\mathbf{H}}_1^*
\tag{3.43}
$$

Then we can calculate the singular value decomposition of $\mathbf{Q}$ and use the same method used in Sect. 3.1 to design the precoders. For the sake of brevity, we do not include the decoding and the proof of full diversity. They are similar in nature to what we presented earlier for users with 4 transmit antennas.

3.4.2 More Receive Antennas, i.e., $M > J = N$

For the case of $M > J = N$, we can pick the $J = N$ receive antennas with the best channel conditions among all M receive antennas for User k at time slot k. In what follows, we illustrate our selection criterion and prove that it provides full diversity. For simplification, we assume $M = 6$, $J = N = 4$.

We assume the channel matrix for User k, $k = 1, \ldots, 4$, is

$$
\mathbf{H}_k = [h_k(i, j)]_{6 \times 4}
\tag{3.44}
$$

Then we pick the 4 rows in $\mathbf{H}_k$ with the largest norms and put them in a matrix $\widetilde{\mathbf{H}}_k$. So we have

$$\|\widetilde{\mathbf{H}}_k\|_F^2 > \frac{2}{3}\|\mathbf{H}_k\|_F^2 \tag{3.45}$$

We just need to pick the 4 antennas corresponding to these 4 rows to finish the antenna selection. Then we can use the scheme proposed in Sect. 3.1 to design the precoder. Next we prove that each user achieves full diversity. We take User 1 for example and start with Equation (3.35), i.e.,

$$P(\mathbf{d} \to \overline{\mathbf{d}}|\widehat{\mathbf{H}}) \le \exp\left(-\frac{\rho \cdot \sum_{i=1}^{4} |h_1^1(i, 1)|^2 |\gamma_1 + k_1\gamma_2 + k_2\gamma_3 + k_3\gamma_4|^2}{4}\right) \tag{3.46}$$

Note that

$$\sum_{i=1}^{4} |h_1^1(i, 1)|^2 = \|\widetilde{\mathbf{H}}_1\mathbf{v}'\|_F^2 > \frac{\|\widetilde{\mathbf{H}}_1\mathbf{V}\|_F^2}{4} = \frac{\|\widetilde{\mathbf{H}}_1\|_F^2}{4} \tag{3.47}$$

where we have used Equation (3.12) and the fact that multiplying by a unitary matrix does not change the norm of a matrix. Then by Equation (3.45), we have

$$\sum_{i=1}^{4} |h_1^1(i, 1)|^2 > \frac{\|\widetilde{\mathbf{H}}_1\|_F^2}{4} > \frac{1}{6}\|\mathbf{H}_1\|_F^2 = \frac{\sum_{i=1}^{4} \sum_{j=1}^{6} |h_1(i, j)|^2}{6} \tag{3.48}$$

Substituting (3.48) in (3.46), we have

$$P(\mathbf{d} \to \overline{\mathbf{d}}|\widehat{\mathbf{H}}) \le \exp\left(-\frac{\rho \cdot \sum_{i=1}^{4} \sum_{j=1}^{6} |h_1(i, j)|^2 |\gamma_1 + k_1\gamma_2 + k_2\gamma_3 + k_3\gamma_4|^2}{24}\right) \tag{3.49}$$

Using the same techniques presented in Sect. 3.1, we have

$$P(\mathbf{d} \to \overline{\mathbf{d}}) < \frac{1}{\prod_{j=1}^{24}[1 + (\rho\zeta'/24)]} \tag{3.50}$$

At high SNRs, one can neglect the one in the denominator and get

$$P(\mathbf{d} \to \overline{\mathbf{d}}) \le \left(\frac{\rho\zeta'}{24}\right)^{-24} \tag{3.51}$$

Then, it is easy to show that the diversity of User 1 is 24, i.e., full diversity. Similarly, we can prove that the diversity of any other user is also full diversity.

Note that although we can achieve full diversity for each user, we only use $J = N$ receive antennas at each time slot. In other words, we do not use all receive antennas. In what follows, we show that besides achieving full diversity, we can further increase the array gain by a simple iterative decoding method.

We let $\begin{pmatrix} \widehat{s_1} \\ \widehat{s_2} \\ \widehat{s_3} \\ \widehat{s_4} \end{pmatrix}$, $\begin{pmatrix} \widehat{t_1} \\ \widehat{t_2} \\ \widehat{t_3} \\ \widehat{t_4} \end{pmatrix}$, $\begin{pmatrix} \widehat{z_1} \\ \widehat{z_2} \\ \widehat{z_3} \\ \widehat{z_4} \end{pmatrix}$, denote the detected signals of Users 2, 3, 4, respec-

tively. We subtract the term $\overline{\mathbf{H}}_2 \begin{pmatrix} \widehat{s_1} \\ \widehat{s_2} \\ \widehat{s_3} \\ \widehat{s_4} \end{pmatrix}$, $\overline{\mathbf{H}}_3 \begin{pmatrix} \widehat{t_1} \\ \widehat{t_2} \\ \widehat{t_3} \\ \widehat{t_4} \end{pmatrix}$, $\overline{\mathbf{H}}_4 \begin{pmatrix} \widehat{z_1} \\ \widehat{z_2} \\ \widehat{z_3} \\ \widehat{z_4} \end{pmatrix}$, from Equation (3.20)

to remove the effect of Users 2, 3, 4 to have

$$\overline{\mathbf{y}} - \sqrt{E_s}\left(\overline{\mathbf{H}}_2 \begin{pmatrix} \widehat{s_1} \\ \widehat{s_2} \\ \widehat{s_3} \\ \widehat{s_4} \end{pmatrix} - \overline{\mathbf{H}}_3 \begin{pmatrix} \widehat{t_1} \\ \widehat{t_2} \\ \widehat{t_3} \\ \widehat{t_4} \end{pmatrix} - \overline{\mathbf{H}}_4 \begin{pmatrix} \widehat{z_1} \\ \widehat{z_2} \\ \widehat{z_3} \\ \widehat{z_4} \end{pmatrix} \right) = \sqrt{E_s}\overline{\mathbf{H}}_1 \begin{pmatrix} c_1 \\ c_2 \\ c_3 \\ c_4 \end{pmatrix} + \overline{\mathbf{n}} + \boldsymbol{\sigma}$$

$$(3.52)$$

where $\boldsymbol{\sigma} = \boldsymbol{\sigma}_1 + \boldsymbol{\sigma}_2 + \boldsymbol{\sigma}_3$ and

$$\boldsymbol{\sigma}_1 = \sqrt{E_s}\overline{\mathbf{H}}_2 \left(\begin{pmatrix} s_1 \\ s_2 \\ s_3 \\ s_4 \end{pmatrix} - \begin{pmatrix} \widehat{s_1} \\ \widehat{s_2} \\ \widehat{s_3} \\ \widehat{s_4} \end{pmatrix} \right),$$

$$\boldsymbol{\sigma}_2 = \overline{\mathbf{H}}_3 \left(\begin{pmatrix} t_1 \\ t_2 \\ t_3 \\ t_4 \end{pmatrix} - \begin{pmatrix} \widehat{t_1} \\ \widehat{t_2} \\ \widehat{t_3} \\ \widehat{t_4} \end{pmatrix} \right),$$

$$\boldsymbol{\sigma}_3 = \overline{\mathbf{H}}_4 \left(\begin{pmatrix} z_1 \\ z_2 \\ z_3 \\ z_4 \end{pmatrix} - \begin{pmatrix} \widehat{z_1} \\ \widehat{z_2} \\ \widehat{z_3} \\ \widehat{z_4} \end{pmatrix} \right)$$

$$(3.53)$$

denote the residual error. Then we can multiply both sides of Equation (3.52) by $\overline{\mathbf{H}}_1^{\dagger}$ and use the same method in Sect. 3.2 to detect the signals of User 1. In what follows, we first show that the method still provides full diversity to User 1. There are two factors that result in an error for User 1. The first one is error in decoding symbols of User 1 after removing the effect of other users and the second one is the error in detecting the symbols of other users at the first time, i.e., error propagation. Let $\Pr(\mathbf{d}_1 \to \overline{\mathbf{d}_1})$ denote the pairwise error probability for User 1, we separate these two events to have

$$\begin{aligned} \Pr(\mathbf{d}_1 \to \overline{\mathbf{d}_1}) &= \Pr\{\mathbf{d}_1 \to \overline{\mathbf{d}_1}|\boldsymbol{\sigma} = 0\}\Pr\{\boldsymbol{\sigma} = 0\} \\ &\quad + \Pr\{\mathbf{d}_1 \to \overline{\mathbf{d}_1}|\boldsymbol{\sigma} \neq 0\}\Pr\{\boldsymbol{\sigma} \neq 0\} \\ &= \Pr\{\mathbf{d}_1 \to \overline{\mathbf{d}_1}|\boldsymbol{\sigma} = 0\}(1 - \Pr\{\boldsymbol{\sigma} \neq 0\}) \\ &\quad + \Pr\{\mathbf{d}_1 \to \overline{\mathbf{d}_1}|\boldsymbol{\sigma} \neq 0\}\Pr\{\boldsymbol{\sigma} \neq 0\} \end{aligned} \qquad (3.54)$$

Since $\Pr\{\mathbf{d}_1 \to \overline{\mathbf{d}_1}|\sigma \neq 0\} \leq 1$ and $1 - \Pr\{\sigma \neq 0\} \leq 1$, we have

$$\Pr(\mathbf{d}_1 \to \overline{\mathbf{d}_1}) \leq \Pr\{\mathbf{d}_1 \to \overline{\mathbf{d}_1}|\sigma = 0\}(1 - \Pr\{\sigma \neq 0\}) + \Pr\{\sigma \neq 0\}$$
$$\leq \Pr\{\mathbf{d}_1 \to \overline{\mathbf{d}_1}|\sigma = 0\} + \Pr\{\sigma \neq 0\} \tag{3.55}$$

Note that when $\sigma = 0$, we can follow the steps in Sect. 3.3 to detect the signals of User 1 and by the same technique used in Sect. 3.3, we can easily derive

$$\Pr\{\mathbf{d}_1 \to \overline{\mathbf{d}_1}|\sigma = 0\} \leq \left(\frac{\rho\zeta'}{24}\right)^{-24} = \tau_1 \rho^{-24} \tag{3.56}$$

where τ_1 is a constant. Further, from (3.51), we know that

$$\Pr\{\sigma \neq 0\} \leq \Pr\{\sigma_1 \neq 0\} + \Pr\{\sigma_2 \neq 0\} + \Pr\{\sigma_3 \neq 0\} \leq \tau_2 \rho^{-24} + \tau_3 \rho^{-24} + \tau_4 \rho^{-24} \tag{3.57}$$

where τ_2, τ_3, τ_4 are all constants. Substituting (3.56) and (3.57) in (3.55), we get

$$\Pr(\mathbf{d}_1 \to \overline{\mathbf{d}_1}) \leq (\tau_1 + \tau_2 + \tau_3 + \tau_4)\rho^{-24} \tag{3.58}$$

Using (3.58), it is easy to show that the diversity $d \geq 24$. Also, it is easy to show that the diversity $d \leq 24$. So the diversity for User 1 is still full diversity. Similarly, we can show that all the other users can also achieve full diversity. In addition, since all the receive antennas are used in the decoding efficiently, it is obvious that the coding gain will be increased.

3.4.3 More Users, i.e., $J > M = N$

In this section, we consider the case that $J > M = N$. For simplification, we assume $J = 6$, $M = N = 4$. First, we assume User k transmits codewords

$$\mathbf{c}_k = \begin{pmatrix} c_{k1} & c_{k2} & c_{k3} & c_{k4} & c_{k5} & c_{k6} \\ c_{k2} & c_{k3} & c_{k4} & c_{k5} & c_{k6} & c_{k1} \\ c_{k3} & c_{k4} & c_{k5} & c_{k6} & c_{k1} & c_{k2} \\ c_{k4} & c_{k5} & c_{k6} & c_{k1} & c_{k2} & c_{k3} \\ c_{k5} & c_{k6} & c_{k1} & c_{k2} & c_{k3} & c_{k4} \\ c_{k6} & c_{k1} & c_{k2} & c_{k3} & c_{k4} & c_{k5} \end{pmatrix} \tag{3.59}$$

Channel matrix and precoder for User k are given by $\mathbf{H}_k = [h_k(i, j)]_{4 \times 4}$ and $\mathbf{A}_k^l = [a_k^l(i, j)]_{4 \times 6}$, respectively, where l denotes the time slot. Note that we can only have four orthogonal directions at most since there are four receive antennas. In order to get the orthogonal structure, we let

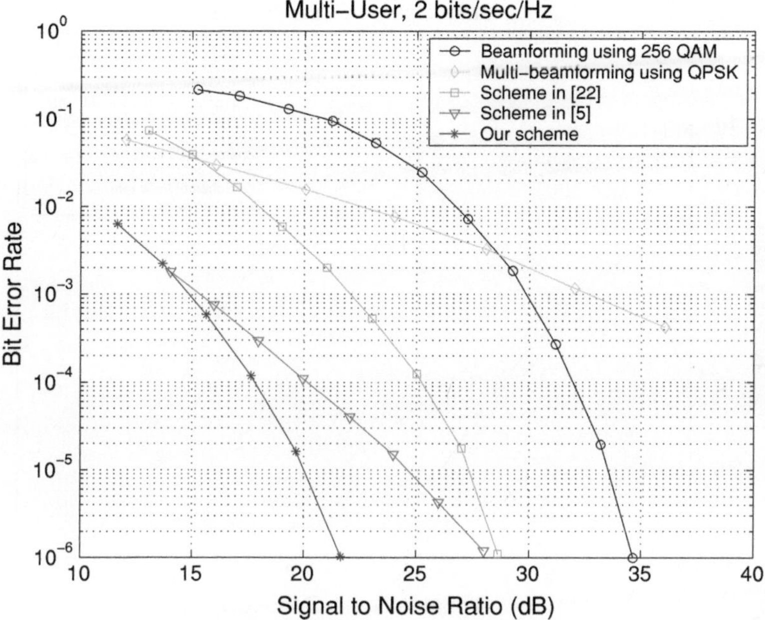

Fig. 3.3 Simulation results for 4 users each with 4 transmit antennas and one receiver with 4 receive antennas

$$\mathbf{A}_5^1 = \mathbf{A}_6^1 = \mathbf{A}_5^2 = \mathbf{A}_6^2 = \mathbf{A}_3^3 = \mathbf{A}_4^3 = \mathbf{A}_3^4 = \mathbf{A}_4^4 = \mathbf{A}_1^5 = \mathbf{A}_2^5 = \mathbf{A}_1^6 = \mathbf{A}_2^6 = \mathbf{0}_{4\times 4}$$
$$(3.60)$$

Then we can use the method proposed in Sect. 3.1 to complete the remaining precoder design. Finally, we can have the following equivalent channel equation

$$\bar{\mathbf{y}} = \sqrt{E_s}\overline{\mathbf{H}}\begin{pmatrix}\mathbf{c}_1\\\mathbf{c}_2\\\mathbf{c}_3\\\mathbf{c}_4\\\mathbf{c}_5\\\mathbf{c}_6\end{pmatrix} + \bar{\mathbf{n}} = \sqrt{E_s}\begin{pmatrix}\overline{\mathbf{H}}_1^1 & \overline{\mathbf{H}}_2^1 & \overline{\mathbf{H}}_3^1 & \overline{\mathbf{H}}_4^1 & \mathbf{0} & \mathbf{0}\\\overline{\mathbf{H}}_1^2 & \overline{\mathbf{H}}_2^2 & \overline{\mathbf{H}}_3^2 & \overline{\mathbf{H}}_4^2 & \mathbf{0} & \mathbf{0}\\\overline{\mathbf{H}}_1^3 & \overline{\mathbf{H}}_2^3 & \mathbf{0} & \mathbf{0} & \overline{\mathbf{H}}_5^3 & \overline{\mathbf{H}}_6^3\\\overline{\mathbf{H}}_1^4 & \overline{\mathbf{H}}_2^4 & \mathbf{0} & \mathbf{0} & \overline{\mathbf{H}}_5^4 & \overline{\mathbf{H}}_6^4\\\mathbf{0} & \mathbf{0} & \overline{\mathbf{H}}_3^5 & \overline{\mathbf{H}}_4^5 & \overline{\mathbf{H}}_5^5 & \overline{\mathbf{H}}_6^5\\\mathbf{0} & \mathbf{0} & \overline{\mathbf{H}}_3^6 & \overline{\mathbf{H}}_4^6 & \overline{\mathbf{H}}_5^6 & \overline{\mathbf{H}}_6^6\end{pmatrix}\begin{pmatrix}\mathbf{c}_1\\\mathbf{c}_2\\\mathbf{c}_3\\\mathbf{c}_4\\\mathbf{c}_5\\\mathbf{c}_6\end{pmatrix} + \bar{\mathbf{n}} \quad (3.61)$$

where the equivalent channel matrix $\overline{\mathbf{H}}$ is a 24×24 matrix, noise vector $\bar{\mathbf{n}}$ is a 24×1 vector and $\overline{\mathbf{H}}_k^l$ denotes the equivalent channel matrix of User k in time slot l. Both $\overline{\mathbf{H}}_k^l$ and $\mathbf{0}$ are 4×4 matrices. Note that with our proposed precoder design in Sect. 3.1, each column in matrix $\overline{\mathbf{H}}$ is orthogonal to all other columns. So we can use the method in Sect. 3.2 to decode the symbols of each user. We can also prove that we can achieve full diversity for each user by the same method used in Sect. 3.3. More specifically, when we prove full diversity for User 1, similar to Equations (3.33), (3.34), we derive

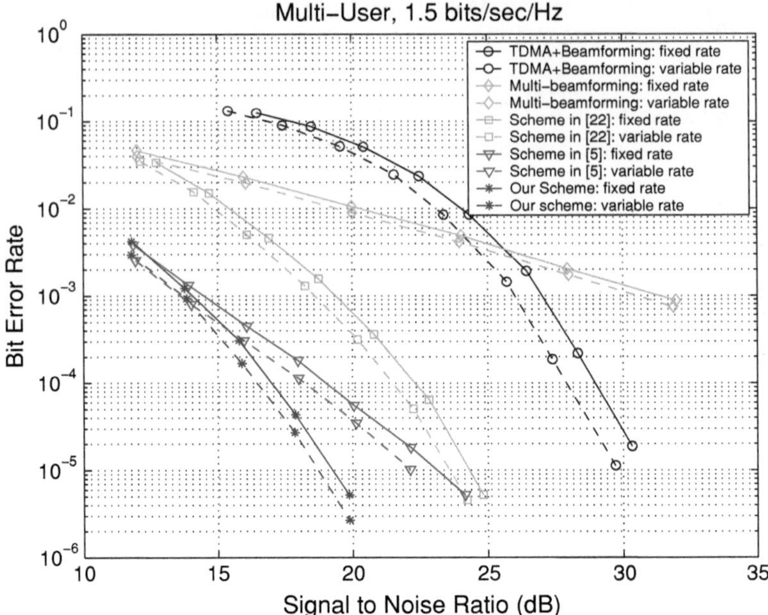

Fig. 3.4 Simulation results for 4 users each with 4 transmit antennas and one receiver with 4 receive antennas when the rate can be adapted

$$P(\mathbf{d} \to \overline{\mathbf{d}}|\widehat{\mathbf{H}}) \leq \exp\left(-\frac{\rho \boldsymbol{\varepsilon}^{\dagger}\mathbf{R}^{\dagger}||\overline{\mathbf{H}}_1||^2\mathbf{R}\boldsymbol{\varepsilon}}{4}\right) = \exp\left(-\frac{\rho\zeta}{4}\right) \qquad (3.62)$$

where

$$\overline{\mathbf{H}}_1 = \left(\overline{\mathbf{H}}_1^1 \ \overline{\mathbf{H}}_1^2 \ \overline{\mathbf{H}}_1^3 \ \overline{\mathbf{H}}_1^4 \ \mathbf{0} \ \mathbf{0}\right)^T \qquad (3.63)$$

$$\zeta = \sum_{i=1}^{4} |h_1^1(i,1)|^2|\gamma_1 + k_1\gamma_2 + k_2\gamma_3 + k_3\gamma_4 + k_4\gamma_5 + k_5\gamma_6|^2$$

$$+ \sum_{i=1}^{4} |h_1^2(i,1)|^2|k_5\gamma_1 + \gamma_2 + k_1\gamma_2 + k_3\gamma_4 + k_3\gamma_5 + k_4\gamma_6|^2$$

$$+ \sum_{i=1}^{4} |h_1^3(i,1)|^2|k_4\gamma_1 + k_5\gamma_2 + \gamma_3 + k_1\gamma_4 + k_2\gamma_5 + k_3\gamma_6|^2$$

$$+ \sum_{i=1}^{4} |h_1^4(i,1)|^2|k_3\gamma_1 + k_4\gamma_2 + k_5\gamma_3 + \gamma_4 + k_1\gamma_5 + k_2\gamma_6|^2 \qquad (3.64)$$

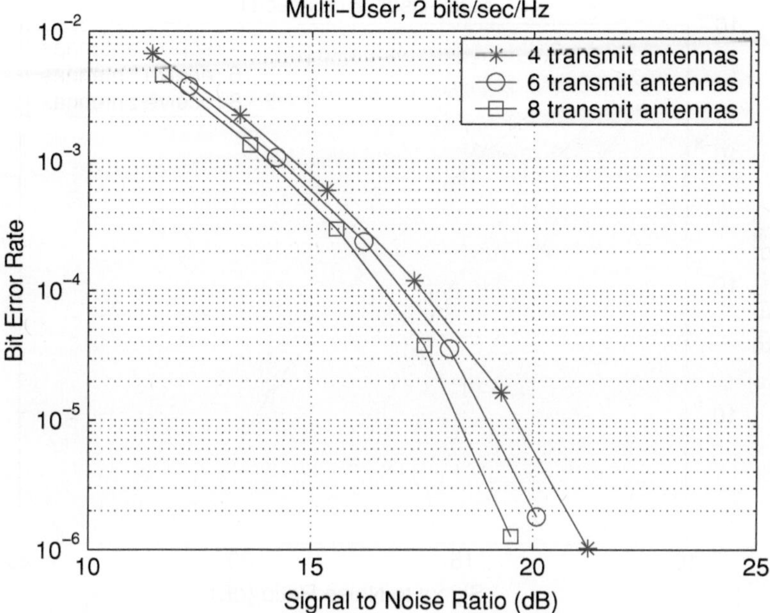

Fig. 3.5 Simulation results for 4 users each with different number of transmit antennas and one receiver with 4 receive antennas

Using the techniques in Sect. 3.3, we show

$$P(\mathbf{d} \to \overline{\mathbf{d}}) \le \left(\frac{\rho\zeta'}{24}\right)^{-24} \tag{3.65}$$

where

$$\zeta' = \frac{|\gamma_1 + k_1\gamma_2 + k_2\gamma_3 + k_3\gamma_4 + k_4\gamma_5 + k_5\gamma_6|^2}{1 + \sum_{j=1}^{5} k_j^2} \tag{3.66}$$

It is easy to see that full diversity for User 1 is achieved. Similarly, we can prove that we can achieve full diversity for other users.

3.5 Simulation Results

In this section, we provide simulation results that confirm our analysis in the previous sections. The performance of our proposed scheme is shown in Figs. 3.3, 3.4, 3.5, 3.6 and 3.7. In Fig. 3.3, we consider 4 users each equipped with 4 transmit antennas and a receiver with 4 receive antennas. We compare our scheme using QPSK and Equations (3.28), (3.29) with the scheme proposed in [4] for the same configuration

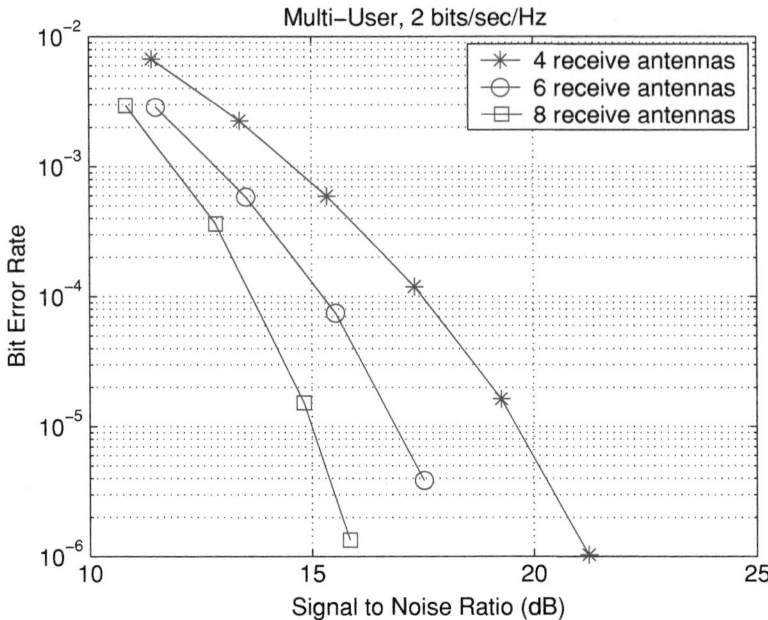

Fig. 3.6 Simulation results for 4 users each with 4 transmit antennas and one receiver with different number of receive antennas

without channel information at the transmitter. With 4 receive antennas, the multi-user detection (MUD) method offered in [4] cancels the interference and provides a diversity of 4. Our scheme also cancels the interference completely but provides a diversity of 16 by utilizing the channel information at the transmitter. We also compare our scheme with two other schemes that can realize interference cancellation and full diversity as well. In both of these two schemes, we assume there are 4 users each with 4 transmit antennas and one receiver with 4 receive antennas. The first scheme is to let only one user transmit using single beamforming at each time slot [5]. In order to have the same bit-rate, 256-QAM is used. The second scheme is to let the first 2 users transmit using the scheme for 2 users in [2] at the first 4 time slots and the second 2 users transmit at the second 4 time slots. Similarly, in order to have the same bit-rate, 16-QAM is used. The results, in Fig. 3.3, show that our scheme outperforms both of these two schemes, by 13 dB and 8 dB, respectively. Finally, we compare our scheme with the following TDMA multiple beamforming scheme. At each time slot, only one user transmits using multiple beamforming, i.e., sends 4 symbols along the 4 eigenvectors of the channel matrix [5]. QPSK is used to match the rate. From the simulation, our scheme outperforms this scheme whose diversity is only 1.

In addition, in Fig. 3.4, we compare our scheme with all mentioned schemes when the rate is changed to adapt with the received SNR. In the simulation, the single beamforming scheme switches between 16-QAM and 256-QAM. The multiple

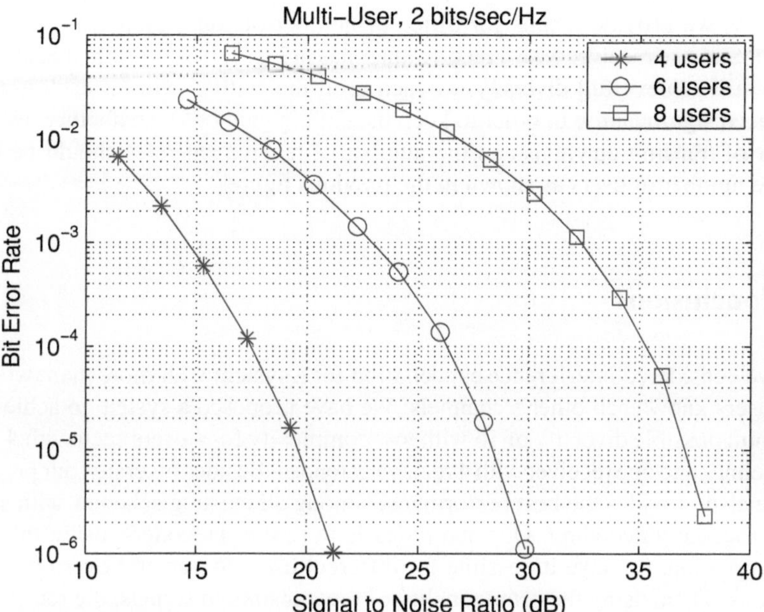

Fig. 3.7 Simulation results for different number of users each with 4 transmit antennas and one receiver with 4 receive antennas

beamforming scheme, the MUD scheme in [4] and our scheme all switch between BPSK and QPSK. The interference cancellation scheme in [2] switches between QPSK and 16-QAM. The threshold to switch between the two rates is properly chosen such that the two constellations are used with the same probability in each scheme. So the rate of all these schemes is 1.5 bits/sec/Hz. We have also provided a "fixed rate" set of simulation results. In all cases, for 1.5 bits/sec/Hz, what we mean by "fixed rate" is the average between the performance of two fixed-rate systems using BPSK and QPSK. From Fig. 3.4, we can see that adapting the rate can improve the performance compared with using a fixed rate. Also we can see that even with variable rate, our scheme provides the best performance.

Next, we present results for 4 users with 4, 6 and 8 transmit antennas and one receiver with 4 receive antennas in Fig. 3.5. When the number of users and the number of receive antennas are fixed, by increasing the number of transmit antennas from 4 to 8, we will have a higher diversity. As we have proved, the diversity is always full diversity using our proposed scheme in Sect. 3.4.

Further, we show the results for 4 users each with 4 transmit antennas and one receiver with 4, 6 and 8 receive antennas in Fig. 3.6. By increasing the number of receive antennas from 4 to 8, the diversity increases from 16 to 32. Therefore, extra receive antennas will provide extra diversity and the resulting diversity of the system is always NM, i.e., full diversity, which confirms our theoretical analysis.

Finally, we provide results for different number of users each with 4 transmit antennas and one receiver with 4 receive antennas in Fig. 3.7. We can see that although we can still achieve full diversity for each user, the coding gain for each user will be reduced significantly. In order to keep the coding gain on a satisfactory level, the number of transmit antennas and the number of receive antennas should be larger than the number of users as shown in the previous figures.

3.6 Conclusions

We have considered interference cancellation for a system with more than two users when users know each other's channels. We have proposed a system to achieve the maximum possible diversity of 16 with low complexity for 4 users each with 4 transmit antennas and one receiver with 4 receive antennas. Besides diversity, our proposed scheme also provides the best performance among all existing schemes with simple array processing decoding. Our main idea is to design precoders, using the channel information, to make it possible for different users to transmit over orthogonal directions. Then, using the orthogonality of the transmitted signals, the receiver can separate them and decode the signals independently. We have analytically proved that the system provides full diversity to each user and extended the results to any number of users each with any number of transmit antennas and one receiver with any number of receive antennas. Further work includes the extension of our scheme to the situation with only limited feedback.

References

1. Jafarkhani, H.: A quasi-orthogonal space-time block code. IEEE Trans. Commun. **49**(1), 1–4 (2001)
2. Li, F., Jafarkhani, H.: Interference cancellation and detection using precoders. Proceedings of IEEE International Conference on Communications, Dresden, Germany, In (2009)
3. Simon, M.K., Alouini, M.-S. (1st ed.): Digital Communication over Fading Channels. Wiley, New York (2000).
4. Kazemitabar, J., Jafarkhani, H.: Multiuser interference cancellation and detection for users with more than two transmit antennas. IEEE Trans. Comm. **56**(4), 574–583 (2008)
5. Sengul, E., Akay, E., Ayanoglu, E.: Diversity analysis of single and multiple beamforming. IEEE Trans. Comm. **54**(6), 990–993 (2006)

Chapter 4
Interference Cancellation for MAC Using Quantized Feedback

4.1 Channel Model

In this chapter, we assume a quasi-static flat Rayleigh fading channel. The path gains are independent complex Gaussian random variables and fixed during the transmission of one block. There are two users each with two transmit antennas and one receiver with two receive antennas.

We assume that the receiver knows the channel information perfectly but only quantized feedback is available at the transmitter. We want to design a scheme to achieve the following two goals using quantized feedback: (i) Canceling the interference at the receiver, i.e., obtaining the interference-free signals for each user at the receiver, (ii) providing full diversity for each user.

In order to achieve these two goals, we propose the following scheme in time slot 1 as shown in Fig. 4.1: First, we assume that Users 1 and 2 transmit codewords $\mathbf{C}$ and $\mathbf{S}$, respectively. And each user can receive K bits of feedback from the receiver. Second, we design a codebook Υ_1 which contains $L_1 = 2^K$ different precoding matrices for User 1 and a codebook Υ_2 which contains $L_2 = 2^K$ different precoding matrices for User 2. Each codebook is shared by its transmitter and the receiver. Also we let $\Upsilon_i[j]$ denote the jth matrix in Codebook Υ_i. Third, the receiver sends back an index ℓ_1 to User 1 using K bits of feedback and an index ℓ_2 to User 2 using another K bits of feedback. Finally, User 1 chooses $\Upsilon_1[\ell_1]$ as its precoder $\mathbf{A}^1$ and transmits the pre-coded signals to the receiver. Also User 2 chooses $\Upsilon_2[\ell_2]$ as its precoder $\mathbf{B}^1$ and transmits the pre-coded signals to the receiver. After receiving the signals from Users 1 and 2, the receiver decodes the signals for each user separately using an array processing method.

In time slot 2, the scheme will be exactly the same as that at time slot 1. But the designed codebooks Υ_1' for User 1 and Υ_2' for User 2 in time slot 2 may be different from the codebooks Υ_1 and Υ_2 in time slot 1. Also the feedback indices ℓ_1' and ℓ_2' in time slot 2 may be different from ℓ_1 and ℓ_2 in time slot 1. As a result, the precoders $\mathbf{A}^2$ for User 1 and $\mathbf{B}^2$ for User 2 in time slot 2 may be different from $\mathbf{A}^1$ and $\mathbf{B}^1$ in time slot 1.

F. Li, *Interference Cancellation Using Space-Time Processing and Precoding Design*,
Signals and Communication Technology, DOI: 10.1007/978-3-642-30712-6_4,
© Springer-Verlag Berlin Heidelberg 2013

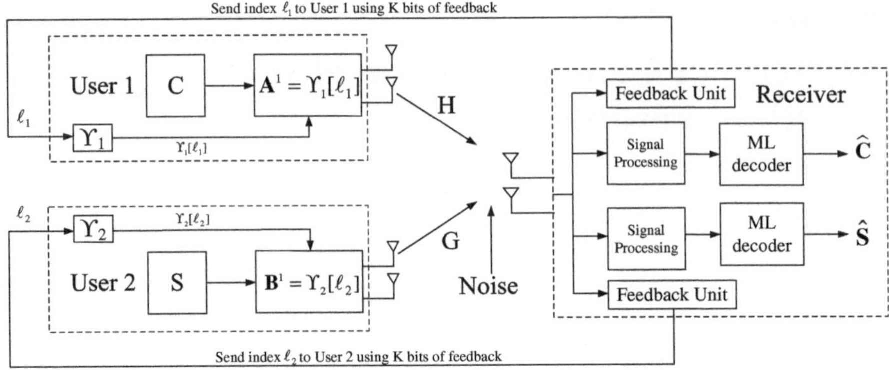

Fig. 4.1 Block Diagram in time slot 1

Now, we demonstrate the input–output relationship of our system. At the first two time slots, the channel matrices for Users 1 and 2 are

$$\mathbf{H} = \begin{pmatrix} h_{11} & h_{12} \\ h_{21} & h_{22} \end{pmatrix}, \quad \mathbf{G} = \begin{pmatrix} g_{11} & g_{12} \\ g_{21} & g_{22} \end{pmatrix} \tag{4.1}$$

respectively, where h_{ij} and g_{ij} are i.i.d. $CN(0, 1)$. For backward compatibility with the case of no feedback in [1], Users 1 and 2 transmit Alamouti codes

$$\mathbf{C} = \begin{pmatrix} c_1 & -c_2^* \\ c_2 & c_1^* \end{pmatrix}, \quad \mathbf{S} = \begin{pmatrix} s_1 & -s_2^* \\ s_2 & s_1^* \end{pmatrix} \tag{4.2}$$

respectively. In order to maximize the diversity and coding gain, we add unitary rotations $\mathbf{R}_1$ and $\mathbf{R}_2$ for codewords of User 1 and User 2, respectively, such that

$$\mathbf{R}_1 \begin{pmatrix} c_1 \\ c_2 \end{pmatrix} = \begin{pmatrix} \tilde{c}_1 \\ \tilde{c}_2 \end{pmatrix}, \quad \mathbf{R}_2 \begin{pmatrix} s_1 \\ s_2 \end{pmatrix} = \begin{pmatrix} \tilde{s}_1 \\ \tilde{s}_2 \end{pmatrix}. \tag{4.3}$$

So the codewords in (4.2) become

$$\tilde{\mathbf{C}} = \begin{pmatrix} \tilde{c}_1 & -\tilde{c}_2^* \\ \tilde{c}_2 & \tilde{c}_1^* \end{pmatrix}, \quad \tilde{\mathbf{S}} = \begin{pmatrix} \tilde{s}_1 & -\tilde{s}_2^* \\ \tilde{s}_2 & \tilde{s}_1^* \end{pmatrix}. \tag{4.4}$$

Let

$$\mathbf{A}^1 = \begin{pmatrix} a_{11}^1 & a_{12}^1 \\ a_{21}^1 & a_{22}^1 \end{pmatrix}, \quad \mathbf{A}^2 = \begin{pmatrix} a_{11}^2 & a_{12}^2 \\ a_{21}^2 & a_{22}^2 \end{pmatrix} \tag{4.5}$$

denote the precoders of User 1 in time slots 1 and 2, respectively. Also,

$$\mathbf{B}^1 = \begin{pmatrix} b_{11}^1 & b_{12}^1 \\ b_{21}^1 & b_{22}^1 \end{pmatrix}, \quad \mathbf{B}^2 = \begin{pmatrix} b_{11}^2 & b_{12}^2 \\ b_{21}^2 & b_{22}^2 \end{pmatrix} \tag{4.6}$$

denote the precoders of User 2 in time slots 1 and 2, respectively. Here $\|\mathbf{A}^i\|_F^2 = \|\mathbf{B}^i\|_F^2 = 1$, $i = 1, 2$, in order to satisfy the normalization conditions [2].

In time slots 1 and 2, the received signals are respectively denoted by

$$\mathbf{y}^1 = \begin{pmatrix} y_1^1 \\ y_2^1 \end{pmatrix}, \quad \mathbf{y}^2 = \begin{pmatrix} y_1^2 \\ y_2^2 \end{pmatrix}. \tag{4.7}$$

Then, in time slot 1, the signal model can be written as

$$\mathbf{y}^1 = \sqrt{E_s}\mathbf{H}\mathbf{A}^1 \begin{pmatrix} \tilde{c}_1 \\ \tilde{c}_2 \end{pmatrix} + \sqrt{E_s}\mathbf{G}\mathbf{B}^1 \begin{pmatrix} \tilde{s}_1 \\ \tilde{s}_2 \end{pmatrix} + \mathbf{W}^1. \tag{4.8}$$

In time slot 2, we have

$$\mathbf{y}^2 = \sqrt{E_s}\mathbf{H}\mathbf{A}^2 \begin{pmatrix} -\tilde{c}_2^* \\ \tilde{c}_1^* \end{pmatrix} + \sqrt{E_s}\mathbf{G}\mathbf{B}^2 \begin{pmatrix} -\tilde{s}_2^* \\ \tilde{s}_1^* \end{pmatrix} + \mathbf{W}^2 \tag{4.9}$$

where E_s denotes the total transmit energy of each user and $\mathbf{W}^1 = \begin{pmatrix} n_1^1 \\ n_2^1 \end{pmatrix}$, $\mathbf{W}^2 = \begin{pmatrix} n_1^2 \\ n_2^2 \end{pmatrix}$ denote the noise at the receiver in time slots 1 and 2, respectively. We assume that $n_1^1, n_2^1, n_1^2, n_2^2$ are i.i.d complex Gaussian noises with mean 0 and variance 1. In order to simplify the notation, we let

$$\widehat{\mathbf{H}}^i = \mathbf{H}\mathbf{A}^i, \text{ i.e., } \begin{pmatrix} \widehat{h}_{11}^i & \widehat{h}_{12}^i \\ \widehat{h}_{21}^i & \widehat{h}_{22}^i \end{pmatrix} = \begin{pmatrix} h_{11}a_{11}^i + h_{12}a_{21}^i & h_{11}a_{12}^i + h_{12}a_{22}^i \\ h_{21}a_{11}^i + h_{22}a_{21}^i & h_{21}a_{12}^i + h_{22}a_{22}^i \end{pmatrix} \tag{4.10}$$

$$\widehat{\mathbf{G}}^i = \mathbf{G}\mathbf{B}^i, \text{ i.e., } \begin{pmatrix} \widehat{g}_{11}^i & \widehat{g}_{12}^i \\ \widehat{g}_{21}^i & \widehat{g}_{22}^i \end{pmatrix} = \begin{pmatrix} g_{11}b_{11}^i + g_{12}b_{21}^i & g_{11}b_{12}^i + g_{12}b_{22}^i \\ g_{21}b_{11}^i + g_{22}b_{21}^i & g_{21}b_{12}^i + g_{22}b_{22}^i \end{pmatrix} \tag{4.11}$$

where $i = 1, 2$. With these new notations, after applying some simple algebra to Eqs. (4.8) and (4.9), we have

$$\begin{pmatrix} y_1^1 \\ y_2^1 \end{pmatrix} = \sqrt{E_s} \begin{pmatrix} \widehat{h}_{11}^1 & \widehat{h}_{12}^1 & \widehat{g}_{11}^1 & \widehat{g}_{12}^1 \\ \widehat{h}_{21}^1 & \widehat{h}_{22}^1 & \widehat{g}_{21}^1 & \widehat{g}_{22}^1 \end{pmatrix} \begin{pmatrix} \tilde{c}_1 \\ \tilde{c}_2 \\ \tilde{s}_1 \\ \tilde{s}_2 \end{pmatrix} + \begin{pmatrix} n_1^1 \\ n_2^1 \end{pmatrix}, \tag{4.12}$$

$$\begin{pmatrix} (y_1^2)^* \\ (y_2^2)^* \end{pmatrix} = \sqrt{E_s} \begin{pmatrix} (\widehat{h}_{12}^2)^* & -(\widehat{h}_{11}^2)^* & (\widehat{g}_{12}^2)^* & -(\widehat{g}_{11}^2)^* \\ (\widehat{h}_{22}^2)^* & -(\widehat{h}_{21}^2)^* & (\widehat{g}_{22}^2)^* & -(\widehat{g}_{21}^2)^* \end{pmatrix} \begin{pmatrix} \widetilde{c}_1 \\ \widetilde{c}_2 \\ \widetilde{s}_1 \\ \widetilde{s}_2 \end{pmatrix} + \begin{pmatrix} (n_1^2)^* \\ (n_2^2)^* \end{pmatrix} \quad (4.13)$$

Equations (4.12) and (4.13) are the input–output relationship of our system at the first two time slots.

4.2 Interference Cancellation Precoding and Decoding

In this section, we will show the property that our codebooks should possess in order to achieve our first goal, i.e., interference cancellation.

4.2.1 Precoding

First, in time slot 1, by Eq. (4.12), $\widetilde{c}_1, \widetilde{c}_2, \widetilde{s}_1, \widetilde{s}_2$ are transmitted along four equivalent channel vectors $\widehat{\mathbf{H}}^1(1)$, $\widehat{\mathbf{H}}^1(2)$, $\widehat{\mathbf{G}}^1(1)$, $\widehat{\mathbf{G}}^1(2)$, respectively. Suppose that we want to remove the signals of User 2, we can find a 2-by-1 complex vector $\mathbf{g}$ satisfying $\mathbf{g}^\dagger \widehat{\mathbf{G}}^1(1) = \mathbf{g}^\dagger \widehat{\mathbf{G}}^1(2) = 0$. Then by simply multiplying both sides of Eq. (4.12) by $\mathbf{g}^\dagger$, we can remove the signals of User 2. This is our basic idea to achieve the interference cancellation.

However, since $\widehat{\mathbf{G}}^1(1)$, $\widehat{\mathbf{G}}^1(2)$ are 2-by-1 complex vectors, a non-zero complex vector $\mathbf{g}_{2 \times 1}$ that satisfies $\mathbf{g}^\dagger \widehat{\mathbf{G}}^1(1) = \mathbf{g}^\dagger \widehat{\mathbf{G}}^1(2) = 0$ does not exist unless $\widehat{\mathbf{G}}^1(1) = \alpha \widehat{\mathbf{G}}^1(2)$, where α is a constant. Therefore, in order to cancel the interference from User 2, we need $\widehat{\mathbf{G}}^1(1) = \alpha \widehat{\mathbf{G}}^1(2)$. To make $\widehat{\mathbf{G}}^1(1) = \alpha \widehat{\mathbf{G}}^1(2)$, our precoders $\mathbf{A}^1$ and $\mathbf{B}^1$ should have the following properties:

$$\mathbf{A}^1(1) = \mathbf{A}^1(2), \quad \mathbf{B}^1(1) = \mathbf{B}^1(2), \quad (4.14)$$

i.e.,

$$\begin{pmatrix} a_{11}^1 \\ a_{21}^1 \end{pmatrix} = \begin{pmatrix} a_{12}^1 \\ a_{22}^1 \end{pmatrix}, \quad \begin{pmatrix} b_{11}^1 \\ b_{21}^1 \end{pmatrix} = \begin{pmatrix} b_{12}^1 \\ b_{22}^1 \end{pmatrix}. \quad (4.15)$$

Since we choose a matrix in the codebook Υ_1 as the precoder for User 1 and a matrix in the codebook Υ_2 as the precoder for User 2, Eq. (4.14) results in:

$$\Upsilon_1[i](1) = \Upsilon_1[i](2), \quad \Upsilon_2[j](1) = \Upsilon_2[j](2), \quad (4.16)$$

i.e., the two columns of any matrix in codebooks Υ_1 and Υ_2 should be the same. From Eqs. (4.10), (4.11), and (4.15), it is easy to see that the resulted $\widehat{\mathbf{G}}^1(1)$, $\widehat{\mathbf{G}}^1(2)$ satisfy $\widehat{\mathbf{G}}^1(1) = \widehat{\mathbf{G}}^1(2)$, i.e.,

$$\begin{pmatrix} \widehat{h}^1_{11} \\ \widehat{h}^1_{21} \end{pmatrix} = \begin{pmatrix} \widehat{h}^1_{12} \\ \widehat{h}^1_{22} \end{pmatrix}, \quad \begin{pmatrix} \widehat{g}^1_{11} \\ \widehat{g}^1_{21} \end{pmatrix} = \begin{pmatrix} \widehat{g}^1_{12} \\ \widehat{g}^1_{22} \end{pmatrix}. \tag{4.17}$$

Then (4.12) can be written as

$$\begin{pmatrix} y^1_1 \\ y^1_2 \end{pmatrix} = \sqrt{E_s} \begin{pmatrix} \widehat{h}^1_{11} & \widehat{h}^1_{11} & \widehat{g}^1_{11} & \widehat{g}^1_{11} \\ \widehat{h}^1_{21} & \widehat{h}^1_{21} & \widehat{g}^1_{21} & \widehat{g}^1_{21} \end{pmatrix} \begin{pmatrix} \widetilde{c}_1 \\ \widetilde{c}_2 \\ \widetilde{s}_1 \\ \widetilde{s}_2 \end{pmatrix} + \begin{pmatrix} n^1_1 \\ n^1_2 \end{pmatrix}. \tag{4.18}$$

Based on Eq. (4.18), it is easy to find a complex vector $\mathbf{g}$ satisfying $\mathbf{g}^{\dagger} \begin{pmatrix} \widehat{g}^1_{11} \\ \widehat{g}^1_{21} \end{pmatrix} = 0$ to remove the signals of User 2. Equation (4.16) represents the property that our codebooks need in order to achieve interference cancellation.

Similarly, in time slot 2, our precoders should satisfy

$$\mathbf{A}^2(1) = \mathbf{A}^2(2), \quad \mathbf{B}^2(1) = \mathbf{B}^2(2). \tag{4.19}$$

Then using the codebook $\varUpsilon'_1$ and $\varUpsilon'_2$, for Users 1 and 2, respectively, any matrix $\varUpsilon'_1[i]$ in the codebook $\varUpsilon'_1$ and any matrix $\varUpsilon'_2[j]$ in the codebook $\varUpsilon'_2$ have the following properties:

$$\varUpsilon'_1[i](1) = \varUpsilon'_1[i](2), \quad \varUpsilon'_2[j](1) = \varUpsilon'_2[j](2). \tag{4.20}$$

Then (4.13) can be written as

$$\begin{pmatrix} (y^2_1)^* \\ (y^2_2)^* \end{pmatrix} = \sqrt{E_s} \begin{pmatrix} (\widehat{h}^2_{12})^* & -(\widehat{h}^2_{12})^* & (\widehat{g}^2_{12})^* & -(\widehat{g}^2_{12})^* \\ (\widehat{h}^2_{22})^* & -(\widehat{h}^2_{22})^* & (\widehat{g}^2_{22})^* & -(\widehat{g}^2_{22})^* \end{pmatrix} \begin{pmatrix} \widetilde{c}_1 \\ \widetilde{c}_2 \\ \widetilde{s}_1 \\ \widetilde{s}_2 \end{pmatrix} + \begin{pmatrix} (n^2_1)^* \\ (n^2_2)^* \end{pmatrix} \tag{4.21}$$

4.2.2 Decoding

In what follows, based on Eqs. (4.18) and (4.21), we illustrate how to cancel the interference of User 2 and decode in detail. First, we introduce some notation to simplify the presentation. In Eqs. (4.18) and (4.21), we let

$$\mathbf{v}^1_h = \begin{pmatrix} \widehat{h}^1_{11} \\ \widehat{h}^1_{21} \end{pmatrix}, \mathbf{v}^1_g = \begin{pmatrix} \widehat{g}^1_{11} \\ \widehat{g}^1_{21} \end{pmatrix}, \bar{\mathbf{y}}^1 = \begin{pmatrix} y^1_1 \\ y^1_2 \end{pmatrix}, \mathbf{n}^1 = \begin{pmatrix} n^1_1 \\ n^1_2 \end{pmatrix} \tag{4.22}$$

$$\mathbf{v}^2_h = \begin{pmatrix} (\widehat{h}^2_{12})^* \\ (\widehat{h}^2_{22})^* \end{pmatrix}, \mathbf{v}^2_g = \begin{pmatrix} (\widehat{g}^2_{12})^* \\ (\widehat{g}^2_{22})^* \end{pmatrix}, \bar{\mathbf{y}}^2 = \begin{pmatrix} (y^2_1)^* \\ (y^2_2)^* \end{pmatrix}, \mathbf{n}^2 = \begin{pmatrix} (n^2_1)^* \\ (n^2_2)^* \end{pmatrix} \tag{4.23}$$

Then we introduce the following complex vectors

$$\bar{\mathbf{v}}_g^1 = \begin{pmatrix} -(\widehat{g}_{21}^1)^* \\ (\widehat{g}_{11}^1)^* \end{pmatrix}, \quad \bar{\mathbf{v}}_g^2 = \begin{pmatrix} -\widehat{g}_{22}^2 \\ \widehat{g}_{12}^2 \end{pmatrix}. \tag{4.24}$$

Note that $\bar{\mathbf{v}}_g^1, \bar{\mathbf{v}}_g^2$ are orthogonal to $\mathbf{v}_g^1, \mathbf{v}_g^2$ in time slots 1 and 2, respectively. In order to cancel the signals from User 2, we can multiply both sides of Eqs. (4.18) and (4.21) by $(\bar{\mathbf{v}}_g^1)^\dagger$ and $(\bar{\mathbf{v}}_g^2)^\dagger$. Then we have

$$(\bar{\mathbf{v}}_g^1)^\dagger \begin{pmatrix} y_1^1 \\ y_2^1 \end{pmatrix} = \sqrt{E_s}(\bar{\mathbf{v}}_g^1)^\dagger \begin{pmatrix} \widehat{h}_{11}^1 & \widehat{h}_{11}^1 \\ \widehat{h}_{21}^1 & \widehat{h}_{21}^1 \end{pmatrix} \begin{pmatrix} \widetilde{c}_1 \\ \widetilde{c}_2 \end{pmatrix} + (\bar{\mathbf{v}}_g^1)^\dagger \begin{pmatrix} n_1^1 \\ n_2^1 \end{pmatrix} \tag{4.25}$$

$$(\bar{\mathbf{v}}_g^2)^\dagger \begin{pmatrix} (y_1^2)^* \\ (y_2^2)^* \end{pmatrix} = \sqrt{E_s}(\bar{\mathbf{v}}_g^2)^\dagger \begin{pmatrix} (\widehat{h}_{12}^2)^* & -(\widehat{h}_{12}^2)^* \\ (\widehat{h}_{22}^2)^* & -(\widehat{h}_{22}^2)^* \end{pmatrix} \begin{pmatrix} \widetilde{c}_1 \\ \widetilde{c}_2 \end{pmatrix} + (\bar{\mathbf{v}}_g^2)^\dagger \begin{pmatrix} (n_1^2)^* \\ (n_2^2)^* \end{pmatrix} \tag{4.26}$$

Now we have removed the signals from User 2. So there is no interference for User 1. The elements of the noise vector $(\bar{\mathbf{v}}_g^1)^\dagger \begin{pmatrix} n_1^1 \\ n_2^1 \end{pmatrix}$ are correlated with covariance $|\bar{\mathbf{v}}_g^1|^2$ and the elements of the noise vector $(\bar{\mathbf{v}}_g^2)^\dagger \begin{pmatrix} (n_1^2)^* \\ (n_2^2)^* \end{pmatrix}$ are correlated with $|\bar{\mathbf{v}}_g^2|^2$. In order to detect the signals of User 1, we need to whiten the noise by multiplying both sides of Eqs. (4.25) and (4.26) by $|\bar{\mathbf{v}}_g^1|^{-1}$ and $|\bar{\mathbf{v}}_g^2|^{-1}$, as follows

$$\frac{(\bar{\mathbf{v}}_g^1)^\dagger}{|\bar{\mathbf{v}}_g^1|} \begin{pmatrix} y_1^1 \\ y_2^1 \end{pmatrix} = \frac{\sqrt{E_s}(\bar{\mathbf{v}}_g^1)^\dagger}{|\bar{\mathbf{v}}_g^1|} \begin{pmatrix} \widehat{h}_{11}^1 & \widehat{h}_{11}^1 \\ \widehat{h}_{21}^1 & \widehat{h}_{21}^1 \end{pmatrix} \begin{pmatrix} \widetilde{c}_1 \\ \widetilde{c}_2 \end{pmatrix} + \frac{(\bar{\mathbf{v}}_g^1)^\dagger}{|\bar{\mathbf{v}}_g^1|} \begin{pmatrix} n_1^1 \\ n_2^1 \end{pmatrix} \tag{4.27}$$

$$\frac{(\bar{\mathbf{v}}_g^2)^\dagger}{|\bar{\mathbf{v}}_g^2|} \begin{pmatrix} (y_1^2)^* \\ (y_2^2)^* \end{pmatrix} = \frac{\sqrt{E_s}(\bar{\mathbf{v}}_g^2)^\dagger}{|\bar{\mathbf{v}}_g^2|} \begin{pmatrix} (\widehat{h}_{12}^2)^* & -(\widehat{h}_{12}^2)^* \\ (\widehat{h}_{22}^2)^* & -(\widehat{h}_{22}^2)^* \end{pmatrix} \begin{pmatrix} \widetilde{c}_1 \\ \widetilde{c}_2 \end{pmatrix} + \frac{(\bar{\mathbf{v}}_g^2)^\dagger}{|\bar{\mathbf{v}}_g^2|} \begin{pmatrix} (n_1^2)^* \\ (n_2^2)^* \end{pmatrix}. \tag{4.28}$$

Using the notation in (4.22), (4.23) and combining Eqs. (4.27) and (4.28), we have

$$\begin{pmatrix} \frac{(\bar{\mathbf{v}}_g^1)^\dagger}{|\bar{\mathbf{v}}_g^1|} \bar{\mathbf{y}}^1 \\ \frac{(\bar{\mathbf{v}}_g^2)^\dagger}{|\bar{\mathbf{v}}_g^2|} \bar{\mathbf{y}}^2 \end{pmatrix} = \sqrt{E_s} \begin{pmatrix} \frac{(\bar{\mathbf{v}}_g^1)^\dagger}{|\bar{\mathbf{v}}_g^1|} \mathbf{v}_h^1 & \frac{(\bar{\mathbf{v}}_g^1)^\dagger}{|\bar{\mathbf{v}}_g^1|} \mathbf{v}_h^1 \\ \frac{(\bar{\mathbf{v}}_g^2)^\dagger}{|\bar{\mathbf{v}}_g^2|} \mathbf{v}_h^2 & -\frac{(\bar{\mathbf{v}}_g^2)^\dagger}{|\bar{\mathbf{v}}_g^2|} \mathbf{v}_h^2 \end{pmatrix} \begin{pmatrix} \widetilde{c}_1 \\ \widetilde{c}_2 \end{pmatrix} + \begin{pmatrix} \frac{(\bar{\mathbf{v}}_g^1)^\dagger}{|\bar{\mathbf{v}}_g^1|} \mathbf{n}^1 \\ \frac{(\bar{\mathbf{v}}_g^2)^\dagger}{|\bar{\mathbf{v}}_g^2|} \mathbf{n}^2 \end{pmatrix}. \tag{4.29}$$

We let $\widehat{\mathbf{H}}$ denote the equivalent channel matrix in (4.29) to simplify the presentation as follows

$$\widehat{\mathbf{H}} = \begin{pmatrix} \widehat{h}_{11} & \widehat{h}_{12} \\ \widehat{h}_{21} & \widehat{h}_{22} \end{pmatrix} = \begin{pmatrix} \widehat{h}_{11} & \widehat{h}_{11} \\ \widehat{h}_{21} & -\widehat{h}_{21} \end{pmatrix} = \begin{pmatrix} \frac{(\bar{\mathbf{v}}_g^1)^\dagger}{|\bar{\mathbf{v}}_g^1|} \mathbf{v}_h^1 & \frac{(\bar{\mathbf{v}}_g^1)^\dagger}{|\bar{\mathbf{v}}_g^1|} \mathbf{v}_h^1 \\ \frac{(\bar{\mathbf{v}}_g^2)^\dagger}{|\bar{\mathbf{v}}_g^2|} \mathbf{v}_h^2 & -\frac{(\bar{\mathbf{v}}_g^2)^\dagger}{|\bar{\mathbf{v}}_g^2|} \mathbf{v}_h^2 \end{pmatrix}. \tag{4.30}$$

Note that $\widehat{\mathbf{H}}$ has the following Single Value Decomposition [3]

$$\widehat{\mathbf{H}} = \mathbf{U}_{\widehat{\mathbf{H}}}\Sigma_{\widehat{\mathbf{H}}}\mathbf{V}_{\widehat{\mathbf{H}}} = \mathbf{U}_{\widehat{\mathbf{H}}}\Sigma_{\widehat{\mathbf{H}}}\begin{pmatrix} \frac{\sqrt{2}}{2} & \frac{\sqrt{2}}{2} \\ \frac{\sqrt{2}}{2} & -\frac{\sqrt{2}}{2} \end{pmatrix} \tag{4.31}$$

where $\mathbf{U}_{\widehat{\mathbf{H}}}$ is a complex matrix and $\Sigma_{\widehat{\mathbf{H}}}$, $\mathbf{V}_{\widehat{\mathbf{H}}}$ are all real matrices. Then we can multiply both sides of Eq. (4.29) by $\mathbf{U}_{\widehat{\mathbf{H}}}^{\dagger}$ as follows

$$\mathbf{U}_{\widehat{\mathbf{H}}}^{\dagger}\begin{pmatrix} \frac{(\overline{\mathbf{v}}_g^1)^{\dagger}}{|\overline{\mathbf{v}}_g^1|}\overline{\mathbf{y}}^1 \\ \frac{(\overline{\mathbf{v}}_g^2)^{\dagger}}{|\overline{\mathbf{v}}_g^2|}\overline{\mathbf{y}}^2 \end{pmatrix} = \sqrt{E_s}\,\mathbf{U}_{\widehat{\mathbf{H}}}^{\dagger}\begin{pmatrix} \frac{(\overline{\mathbf{v}}_g^1)^{\dagger}}{|\overline{\mathbf{v}}_g^1|}\mathbf{v}_h^1 & \frac{(\overline{\mathbf{v}}_g^1)^{\dagger}}{|\overline{\mathbf{v}}_g^1|}\mathbf{v}_h^1 \\ \frac{(\overline{\mathbf{v}}_g^2)^{\dagger}}{|\overline{\mathbf{v}}_g^2|}\mathbf{v}_h^2 & -\frac{(\overline{\mathbf{v}}_g^2)^{\dagger}}{|\overline{\mathbf{v}}_g^2|}\mathbf{v}_h^2 \end{pmatrix}\begin{pmatrix} \widetilde{c}_1 \\ \widetilde{c}_2 \end{pmatrix}$$

$$+ \mathbf{U}_{\widehat{\mathbf{H}}}^{\dagger}\begin{pmatrix} \frac{(\overline{\mathbf{v}}_g^1)^{\dagger}}{|\overline{\mathbf{v}}_g^1|}\mathbf{n}^1 \\ \frac{(\overline{\mathbf{v}}_g^2)^{\dagger}}{|\overline{\mathbf{v}}_g^2|}\mathbf{n}^2 \end{pmatrix}. \tag{4.32}$$

In the above equation, $\mathbf{U}_{\widehat{\mathbf{H}}}^{\dagger}\begin{pmatrix} \frac{(\overline{\mathbf{v}}_g^1)^{\dagger}}{|\overline{\mathbf{v}}_g^1|}\mathbf{n}^1 \\ \frac{(\overline{\mathbf{v}}_g^2)^{\dagger}}{|\overline{\mathbf{v}}_g^2|}\mathbf{n}^2 \end{pmatrix}$ is still white noise and $\mathbf{U}_{\widehat{\mathbf{H}}}^{\dagger}\begin{pmatrix} \frac{(\overline{\mathbf{v}}_g^1)^{\dagger}}{|\overline{\mathbf{v}}_g^1|}\mathbf{v}_h^1 & \frac{(\overline{\mathbf{v}}_g^1)^{\dagger}}{|\overline{\mathbf{v}}_g^1|}\mathbf{v}_h^1 \\ \frac{(\overline{\mathbf{v}}_g^2)^{\dagger}}{|\overline{\mathbf{v}}_g^2|}\mathbf{v}_h^2 & -\frac{(\overline{\mathbf{v}}_g^2)^{\dagger}}{|\overline{\mathbf{v}}_g^2|}\mathbf{v}_h^2 \end{pmatrix}$

is real matrix. So if QAM is used, then we have

$$\mathrm{Real}\left\{\mathbf{U}_{\widehat{\mathbf{H}}}^{\dagger}\begin{pmatrix} \frac{(\overline{\mathbf{v}}_g^1)^{\dagger}}{|\overline{\mathbf{v}}_g^1|}\overline{\mathbf{y}}^1 \\ \frac{(\overline{\mathbf{v}}_g^2)^{\dagger}}{|\overline{\mathbf{v}}_g^2|}\overline{\mathbf{y}}^2 \end{pmatrix}\right\} = \sqrt{E_s}\,\mathbf{U}_{\widehat{\mathbf{H}}}^{\dagger}\begin{pmatrix} \frac{(\overline{\mathbf{v}}_g^1)^{\dagger}}{|\overline{\mathbf{v}}_g^1|}\mathbf{v}_h^1 & \frac{(\overline{\mathbf{v}}_g^1)^{\dagger}}{|\overline{\mathbf{v}}_g^1|}\mathbf{v}_h^1 \\ \frac{(\overline{\mathbf{v}}_g^2)^{\dagger}}{|\overline{\mathbf{v}}_g^2|}\mathbf{v}_h^2 & -\frac{(\overline{\mathbf{v}}_g^2)^{\dagger}}{|\overline{\mathbf{v}}_g^2|}\mathbf{v}_h^2 \end{pmatrix}$$

$$\cdot\, \mathrm{Real}\left\{\begin{pmatrix} \widetilde{c}_1 \\ \widetilde{c}_2 \end{pmatrix}\right\} + \mathrm{Real}\left\{\mathbf{U}_{\widehat{\mathbf{H}}}^{\dagger}\begin{pmatrix} \frac{(\overline{\mathbf{v}}_g^1)^{\dagger}}{|\overline{\mathbf{v}}_g^1|}\mathbf{n}^1 \\ \frac{(\overline{\mathbf{v}}_g^2)^{\dagger}}{|\overline{\mathbf{v}}_g^2|}\mathbf{n}^2 \end{pmatrix}\right\} \tag{4.33}$$

$$\mathrm{Imag}\left\{\mathbf{U}_{\widehat{\mathbf{H}}}^{\dagger}\begin{pmatrix} \frac{(\overline{\mathbf{v}}_g^1)^{\dagger}}{|\overline{\mathbf{v}}_g^1|}\overline{\mathbf{y}}^1 \\ \frac{(\overline{\mathbf{v}}_g^2)^{\dagger}}{|\overline{\mathbf{v}}_g^2|}\overline{\mathbf{y}}^2 \end{pmatrix}\right\} = \sqrt{E_s}\,\mathbf{U}_{\widehat{\mathbf{H}}}^{\dagger}\begin{pmatrix} \frac{(\overline{\mathbf{v}}_g^1)^{\dagger}}{|\overline{\mathbf{v}}_g^1|}\mathbf{v}_h^1 & \frac{(\overline{\mathbf{v}}_g^1)^{\dagger}}{|\overline{\mathbf{v}}_g^1|}\mathbf{v}_h^1 \\ \frac{(\overline{\mathbf{v}}_g^2)^{\dagger}}{|\overline{\mathbf{v}}_g^2|}\mathbf{v}_h^2 & -\frac{(\overline{\mathbf{v}}_g^2)^{\dagger}}{|\overline{\mathbf{v}}_g^2|}\mathbf{v}_h^2 \end{pmatrix}$$

$$\cdot\, \mathrm{Imag}\left\{\begin{pmatrix} \widetilde{c}_1 \\ \widetilde{c}_2 \end{pmatrix}\right\} + \mathrm{Imag}\left\{\mathbf{U}_{\widehat{\mathbf{H}}}^{\dagger}\begin{pmatrix} \frac{(\overline{\mathbf{v}}_g^1)^{\dagger}}{|\overline{\mathbf{v}}_g^1|}\mathbf{n}^1 \\ \frac{(\overline{\mathbf{v}}_g^2)^{\dagger}}{|\overline{\mathbf{v}}_g^2|}\mathbf{n}^2 \end{pmatrix}\right\} \tag{4.34}$$

Therefore, we can use the Maximum-Likelihood method to decode the real parts and imaginary parts of $\widetilde{c}_1$, $\widetilde{c}_2$ separately. For example, when we detect the real parts of $\widetilde{c}_1$, $\widetilde{c}_2$, we have

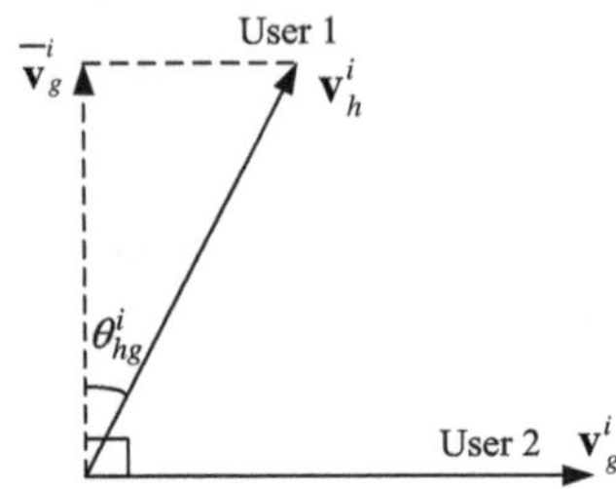

Fig. 4.2 Precoder design illustration

$$
\begin{aligned}
\mathrm{Real}\{\widehat{c}_1, \widehat{c}_2\} = \arg\min_{\mathrm{Real}\{c_1, c_2\}} &\left\| \mathrm{Real}\left\{ \mathbf{U}_{\widehat{\mathbf{H}}}^\dagger \begin{pmatrix} \frac{(\overline{\mathbf{v}}_g^1)^\dagger}{|\overline{\mathbf{v}}_g^1|}\overline{\mathbf{y}}^1 \\ \frac{(\overline{\mathbf{v}}_g^2)^\dagger}{|\overline{\mathbf{v}}_g^2|}\overline{\mathbf{y}}^2 \end{pmatrix} \right\} \right. \\
&\left. -\sqrt{E_s}\mathbf{U}_{\widehat{\mathbf{H}}}^\dagger \begin{pmatrix} \frac{(\overline{\mathbf{v}}_g^1)^\dagger}{|\overline{\mathbf{v}}_g^1|}\mathbf{v}_h^1 & \frac{(\overline{\mathbf{v}}_g^1)^\dagger}{|\overline{\mathbf{v}}_g^1|}\mathbf{v}_h^1 \\ \frac{(\overline{\mathbf{v}}_g^2)^\dagger}{|\overline{\mathbf{v}}_g^2|}\mathbf{v}_h^2 & -\frac{(\overline{\mathbf{v}}_g^2)^\dagger}{|\overline{\mathbf{v}}_g^2|}\mathbf{v}_h^2 \end{pmatrix} \mathrm{Real}\left\{ \begin{pmatrix} \widetilde{c}_1 \\ \widetilde{c}_2 \end{pmatrix} \right\} \right\|_F^2 .
\end{aligned}
\tag{4.35}
$$

Similarly, we can decode the imaginary parts of $\widetilde{c}_1$, $\widetilde{c}_2$, and the signals of User 2. Note that the decoding complexity is symbol-by-symbol.

Till now, we have presented our precoding, decoding methods, and some necessary properties needed by our codebooks to cancel interference for each user. Note that in order to achieve interference cancellation, the only properties needed by our codebooks are (4.16) and (4.20). The remaining degrees of freedom will be used to maximize diversity and coding gain as discussed in the next two sections.

4.3 Feedback Design and Diversity Analysis

In this section, we first propose our feedback scheme, i.e., how to choose an index l_i and send it back to User i. Then we prove that our feedback scheme can achieve full diversity when our codebooks satisfy some conditions.

4.3.1 Feedback Design

First, as illustrated in Fig. 4.2, we define

$$
\cos\theta_{hg}^1 =\ <\mathbf{v}_h^1, \overline{\mathbf{v}}_g^1> =\ \frac{|(\overline{\mathbf{v}}_g^1)^\dagger \mathbf{v}_h^1|}{|\overline{\mathbf{v}}_g^1| \cdot |\mathbf{v}_h^1|}
\tag{4.36}
$$

$$\cos \theta_{hg}^2 = < \mathbf{v}_h^2, \overline{\mathbf{v}}_g^2 > = \frac{|(\overline{\mathbf{v}}_g^2)^\dagger \mathbf{v}_h^2|}{|\overline{\mathbf{v}}_g^2| \cdot |\mathbf{v}_h^2|}. \tag{4.37}$$

Note that the maximum value of $\cos \theta_{hg}^i$ is 1 and the corresponding $\theta_{hg}^i = 0$, which means $\mathbf{v}_h^i$ and $\mathbf{v}_g^i$ are orthogonal to each other.

Now we introduce our feedback scheme with the assumption that User 1 has already got a codebook Υ_1 in time slot 1 and a codebook Υ_1' in time slot 2. Also User 2 has already got codebooks Υ_2 and Υ_2' in time slots 1 and 2, respectively. All these codebooks should possess the property given by (4.16) and (4.20). In time slot 1, the receiver selects an index ℓ_1 within the range from 0 to $L_1 - 1$ and sends it back to User 1. The selection criterion is that with such an index ℓ_1, $|\mathbf{v}_h^1|$ is maximized, where $|\mathbf{v}_h^1| = |\mathbf{HA}^1(1)|$ as given by (4.22) and $\mathbf{A}^1 = \Upsilon_1[\ell_1]$. Maximizing $|\mathbf{v}_h^1|$ is equivalent to maximizing the received SINR for User 1. Therefore, full diversity is also achieved, as shown later. At the same time slot, the receiver also picks an index ℓ_2 and sends it back to User 2. The selection criterion is that with such an index ℓ_2, θ_{hg}^1 is minimized, where θ_{hg}^1 is given by (4.36) in which $\overline{\mathbf{v}}_g^1 = \begin{pmatrix} -g_{21}^* & -g_{22}^* \\ g_{11}^* & g_{12}^* \end{pmatrix} \mathbf{B}^1(1)^*$ as given by (4.24), $\mathbf{B}^1 = \Upsilon_2[\ell_2]$. We will show that by doing so, we can also maximize coding gain within our system framework.

Similarly, in time slot 2, the receiver finds an index ℓ_2' and sends it back to User 2. The selection criterion is that with such an index ℓ_2', $|\mathbf{v}_g^2|$ is maximized. The receiver also finds an index ℓ_1' and sends it back to User 1. The selection criterion is that with such an index ℓ_1', θ_{hg}^2 is minimized.

4.3.2 Diversity Analysis

In what follows, we show that by the above proposed scheme, the diversity for each user is full as long as our codebooks satisfy some conditions. The diversity is defined as

$$d = -\lim_{\rho \to \infty} \frac{\log P_e}{\log \rho} \tag{4.38}$$

where ρ denotes the SNR and P_e represents the probability of error. We first consider Eq. (4.29) to analyze the diversity for User 1. We know $\begin{pmatrix} \widetilde{c}_1 \\ \widetilde{c}_2 \end{pmatrix} = \mathbf{R}_1 \begin{pmatrix} 1 \\ c_2 \end{pmatrix}$ and we define the error matrix $\boldsymbol{\varepsilon} = \begin{pmatrix} 1 \\ c_2 \end{pmatrix} - \begin{pmatrix} \widehat{c}_1 \\ \widehat{c}_2 \end{pmatrix}$. By (4.29), the pairwise error probability (PEP) can be given by the Gaussian tail function as [4]

$$P(\mathbf{d} \to \overline{\mathbf{d}}|\widehat{\mathbf{H}}) = Q\left(\sqrt{\frac{\rho \|\widehat{\mathbf{H}}\mathbf{R}_1\boldsymbol{\varepsilon}\|_F^2}{4}}\right)$$

$$= Q\left(\sqrt{\frac{\rho \boldsymbol{\varepsilon}^\dagger \mathbf{R}_1^\dagger (\widehat{\mathbf{H}})^\dagger \widehat{\mathbf{H}} \mathbf{R}_1 \boldsymbol{\varepsilon}}{4}}\right)$$

$$\leq \exp\left(-\frac{\rho \boldsymbol{\varepsilon}^\dagger \mathbf{R}_1^\dagger (\widehat{\mathbf{H}})^\dagger \widehat{\mathbf{H}} \mathbf{R}_1 \boldsymbol{\varepsilon}}{8}\right) \tag{4.39}$$

where we have used the inequality $Q(x) \leq \exp(-\frac{x^2}{2})$. Now we assume $\mathbf{R}_1\boldsymbol{\varepsilon} = \begin{pmatrix} \gamma_1 \\ \gamma_2 \end{pmatrix}$. Substituting $\mathbf{R}_1\boldsymbol{\varepsilon}$ and $\widehat{\mathbf{H}}$ from Eq. (4.30) in (4.39), we have

$$P(\mathbf{d} \to \overline{\mathbf{d}}|\widehat{\mathbf{H}}) \leq \exp\left(-\frac{\rho}{8}\left(|\widehat{h}_{11}|^2|\gamma_1 + \gamma_2|^2 + |\widehat{h}_{21}|^2|\gamma_1 - \gamma_2|^2\right)\right)$$

$$= \exp\left(-\frac{\rho}{8}\left(\left|\frac{(\overline{\mathbf{v}}_g^1)^\dagger}{|\overline{\mathbf{v}}_g^1|}\mathbf{v}_h^1\right|^2 |\gamma_1 + \gamma_2|^2 + \left|\frac{(\overline{\mathbf{v}}_g^2)^\dagger}{|\overline{\mathbf{v}}_g^2|}\mathbf{v}_h^2\right|^2 |\gamma_1 - \gamma_2|^2\right)\right)$$

$$\leq \exp\left(-\frac{\rho\left(\left|\frac{(\overline{\mathbf{v}}_g^1)^\dagger}{|\overline{\mathbf{v}}_g^1|}\mathbf{v}_h^1\right|^2 |\gamma_1 + \gamma_2|^2\right)}{8}\right). \tag{4.40}$$

Let us define

$$\Delta = \left|\frac{(\overline{\mathbf{v}}_g^1)^\dagger}{|\overline{\mathbf{v}}_g^1|}\mathbf{v}_h^1\right|^2 = \frac{|(\overline{\mathbf{v}}_g^1)^\dagger \mathbf{v}_h^1|^2}{|\overline{\mathbf{v}}_g^1|^2}. \tag{4.41}$$

Using (4.36), we can rewrite Δ as

$$\Delta = |\cos\theta_{hg}^1|^2 \cdot |\mathbf{v}_h^1|^2. \tag{4.42}$$

Substituting (4.42) in (4.40), we have

$$P(\mathbf{d} \to \overline{\mathbf{d}}|\widehat{\mathbf{H}}) \leq \exp\left(-\frac{\rho(|\cos\theta_{hg}^1|^2 \cdot |\mathbf{v}_h^1|^2|\gamma_1 + \gamma_2|^2)}{8}\right). \tag{4.43}$$

Since we choose our precoder $\mathbf{A}^1$ from the codebook Υ_1 such that $|\mathbf{v}_h^1|^2$ is maximized, it is easy to see

$$|\mathbf{v}_h^1|^2 = |\mathbf{H}\mathbf{A}^1(1)|^2 \geq \frac{|\mathbf{H}\overline{\Upsilon}_1|^2}{L} \tag{4.44}$$

where $\overline{\Upsilon}_1$ is a matrix satisfying $\overline{\Upsilon}_1(i) = \Upsilon_1[i](1), i = 1, \ldots, L$, i.e., the ith column of matrix $\overline{\Upsilon}_1$ is the same as the first column of the ith matrix in the codebook Υ_1. We assume $\overline{\Upsilon}_1$ has the following Singular Value Decomposition

$$\overline{\varUpsilon}_1 = \mathbf{U}_{\overline{\varUpsilon}_1} \Sigma_{\overline{\varUpsilon}_1} V_{\overline{\varUpsilon}_1}^{\dagger} = \mathbf{U}_{\overline{\varUpsilon}_1} \begin{pmatrix} \lambda_1^{\overline{\varUpsilon}_1} & 0 \\ 0 & \lambda_2^{\overline{\varUpsilon}_1} \end{pmatrix} V_{\overline{\varUpsilon}_1}^{\dagger}. \tag{4.45}$$

Then (4.44) becomes

$$|\mathbf{v}_h^1|^2 \geq \frac{|\mathbf{HU}_{\overline{\varUpsilon}_1} \Sigma_{\overline{\varUpsilon}_1} V_{\overline{\varUpsilon}_1}^{\dagger}|^2}{L} = \frac{|\lambda_1^{\overline{\varUpsilon}_1}|^2(|h'_{11}|^2 + |h'_{21}|^2) + |\lambda_2^{\overline{\varUpsilon}_1}|^2(|h'_{12}|^2 + |h'_{22}|^2)}{L}$$
$$\tag{4.46}$$

where

$$\mathbf{HU}_{\overline{\varUpsilon}_1} = \begin{pmatrix} h'_{11} & h'_{12} \\ h'_{21} & h'_{22} \end{pmatrix}. \tag{4.47}$$

Since the unitary matrix $\mathbf{U}_{\overline{\varUpsilon}_1}$ does not change the distribution of $\mathbf{H}$, each element of $\mathbf{HU}_{\overline{\varUpsilon}_1}$, i.e., h'_{ij}, is also a Gaussian distributed random variable with mean 0 and variance 1. As a result, (4.43) can be written as

$$P(\mathbf{d} \to \overline{\mathbf{d}} | \widehat{\mathbf{H}})$$
$$\leq \exp\left(-\frac{\rho}{4L}\left(|\cos\theta_{hg}^1|^2(|\lambda_1^{\overline{\varUpsilon}_1}|^2(|h'_{11}|^2 + |h'_{21}|^2)\right.\right.$$
$$\left.\left. +|\lambda_2^{\overline{\varUpsilon}_1}|^2(|h'_{12}|^2 + |h'_{22}|^2))|\gamma_1 + \gamma_2|^2\right)\right). \tag{4.48}$$

Further, we have

$$P(\mathbf{d} \to \overline{\mathbf{d}})$$
$$\leq E\left[\exp\left(-\frac{\rho}{4L}(|\cos\theta_{hg}^1|^2(|\lambda_1^{\overline{\varUpsilon}_1}|^2(|h'_{11}|^2 + |h'_{21}|^2)\right.\right.$$
$$\left.\left. +|\lambda_2^{\overline{\varUpsilon}_1}|^2(|h'_{12}|^2 + |h'_{22}|^2))|\gamma_1 + \gamma_2|^2)\right)\right]$$
$$= E\left[E\left[\exp\left(-\frac{\rho}{4L}(|\cos\theta_{hg}^1|^2(|\lambda_1^{\overline{\varUpsilon}_1}|^2(|h'_{11}|^2 + |h'_{21}|^2)\right.\right.\right.$$
$$\left.\left. +|\lambda_2^{\overline{\varUpsilon}_1}|^2(|h'_{12}|^2 + |h'_{22}|^2))|\gamma_1 + \gamma_2|^2)\right) |\theta_{hg}^1\right]\right]$$
$$\leq E\left[\frac{1}{\prod_{j=1}^2 [1 + (\frac{\rho}{8L}|\cos\theta_{hg}^1|^2|\lambda_j^{\overline{\varUpsilon}_1}|^2|\gamma_1 + \gamma_2|^2)]^2}\right] \tag{4.49}$$

At high SNRs, one can neglect the one in the denominator and get

$$P(\mathbf{d} \to \overline{\mathbf{d}}) \leq \left(\frac{\rho}{8L}\right)^{-4} \prod_{j=1}^2 (|\lambda_j^{\overline{\varUpsilon}_1}| \cdot |\gamma_1 + \gamma_2|)^{-4} E\left[\frac{1}{|\cos\theta_{hg}^1|^8}\right]. \tag{4.50}$$

From (4.50), it is easy to see the diversity for User 1 is 4, full diversity, as long as $\lambda_j^{\overline{\Upsilon}_1} \neq 0$. Note that matrix $\overline{\Upsilon}_1$ is a 2-by-L matrix, where L is the number of matrices in codebook Υ_1. So in order to make $\lambda_j^{\overline{\Upsilon}_1} \neq 0$, we need

1. $L \geq 2$, where L is the number of matrices in our codebook.
2. The rank of matrix $\overline{\Upsilon}_1$ is 2.

Condition 1 requires that $K \geq 1$, where K is the number of feedback bits available to each user. Condition 2 is a constraint we need to design our codebook Υ_1. There is no other constraint on the codebook Υ_1 in order to achieve full diversity. In time slot 1, there is no further requirement on Codebook Υ_2 for User 2 other than (4.16). In time slot 2, by a similar proof, the codebook Υ_2' for User 2 should satisfy the above two conditions and the only requirement on Codebook Υ_1' for User 1 is (4.20). Similarly, we can prove that the diversity for User 2 is also full.

4.4 Coding Gain Analysis and Codebook Design

In the last two sections, we have presented some properties needed by our codebooks in order to achieve interference cancellation and full diversity. However, there are still some degrees of freedom in our codebook design. In this section, we use the remaining degrees of freedom to maximize the coding gain.

By (4.43), in order to maximize coding gain, we need to maximize $|v_h^1|$ and $|\cos \theta_{hg}^1|$. We first analyze v_h^1. Note that

$$v_h^1 = \mathbf{H}\mathbf{A}^1(1). \tag{4.51}$$

To maximize $|v_h^1|$, the best choice for $\mathbf{A}^1(1)$ is [5]

$$\mathbf{A}^1(1) = \frac{1}{\sqrt{2}}\mathbf{V_H}(1) \tag{4.52}$$

where $\mathbf{V_H}$ comes from the singular value decomposition

$$\mathbf{H} = \mathbf{U_H}\Sigma_H\mathbf{V}_H^\dagger = \mathbf{U_H}\begin{pmatrix} \lambda_1 & 0 \\ 0 & \lambda_2 \end{pmatrix}\mathbf{V_H}^\dagger. \tag{4.53}$$

$\mathbf{V_H}(1)$ is the singular vector of $\mathbf{H}$ corresponding to the largest singular value and we assume $\lambda_1 > \lambda_2$ without loss of generality. If we have perfect feedback, we can simply choose $\mathbf{A}^1(1) = \frac{1}{\sqrt{2}}\mathbf{V_H}(1)$ and the precoder $\mathbf{A}^1 = \frac{1}{\sqrt{2}}[\mathbf{V_H}(1), \mathbf{V_H}(1)]$. Since we only have access to quantized feedback, we should design a codebook in which we can find a matrix whose column is the best approximation to $\frac{1}{\sqrt{2}}\mathbf{V_H}(1)$.

It has been shown in [6] that $\mathbf{V_H}(1)$ is an isotropically distributed unitary vector. The intuitive meaning of an isotropically distributed complex unit vector is that it is equally likely to point in any direction in complex space. Therefore, the problem to design a codebook to maximize $|\mathbf{v}_h^1|$ becomes how to pack one-dimensional subspaces of a complex space known as Grassmannian line packing [7]. In other words, it is the problem of finding a set of L_1 one-dimensional subspaces in the complex space that maximize the minimum distance between any pair of subspaces in the set.

The problem of finding optimal line packings using analytical or numerical methods is not new [7–10]. We utilize the existing methodologies in the literature to design a codebook for User 1 in time slot 1.

Now we summarize the procedures to construct our codebook for User 1 in time slot 1:

1. For K bits of feedback, find $L_1 = 2^K$ two-by-one unit norm complex vectors which can maximize the minimum distance between any pair of vectors in the two-dimensional complex space. We denote all these vectors as $\psi_i, i = 1, \ldots, L_1$.
2. Create a codebook Υ_1 that contains $L_1 = 2^K$ matrices satisfying $\Upsilon_1[i] = \frac{1}{\sqrt{2}}[\psi_i, \psi_i]$.

It is easy to check that the created codebook satisfies all the conditions we need. Therefore, $|\mathbf{v}_h^1|$ can be maximized if User 1 adopts the above codebook.

In what follows, we will show that if User 2 adopts the above codebook, $|\cos\theta_{hg}^1|$ will also be maximized. By (4.36), we know that once $|\mathbf{v}_h^1|$ and $|\cos\theta_{hg}^1|$ are maximized at the same time, the coding gain will be maximized. Therefore, the above codebook is the optimal codebook that both User 1 and User 2 should adopt in time slot 1.

First, note that in order to maximize $|\cos\theta_{hg}^1|$, by (4.36), we need $\bar{\mathbf{v}}_g^1 = \eta\mathbf{v}_h^1$, i.e.,

$$\begin{pmatrix} -(\widehat{g}_{21}^1)^* \\ (\widehat{g}_{11}^1)^* \end{pmatrix} = \eta \begin{pmatrix} \widehat{h}_{11}^1 \\ \widehat{h}_{21}^1 \end{pmatrix} \text{ or } \eta \begin{pmatrix} (\widehat{h}_{21}^1)^* \\ -(\widehat{h}_{11}^1)^* \end{pmatrix} = \begin{pmatrix} \widehat{g}_{11}^1 \\ \widehat{g}_{21}^1 \end{pmatrix} \tag{4.54}$$

where η is a constant. Further, we have

$$\eta \begin{pmatrix} (\widehat{h}_{21}^1)^* \\ -(\widehat{h}_{11}^1)^* \end{pmatrix} = \begin{pmatrix} g_{11} & g_{12} \\ g_{21} & g_{22} \end{pmatrix} \begin{pmatrix} b_{11}^1 \\ b_{21}^1 \end{pmatrix}$$

$$\text{or } \begin{pmatrix} b_{11}^1 \\ b_{21}^1 \end{pmatrix} = \eta \begin{pmatrix} g_{11} & g_{12} \\ g_{21} & g_{22} \end{pmatrix}^{-1} \begin{pmatrix} (\widehat{h}_{21}^1)^* \\ -(\widehat{h}_{11}^1)^* \end{pmatrix}. \tag{4.55}$$

Since the norm of $\begin{pmatrix} b_{11}^1 \\ b_{21}^1 \end{pmatrix}$ is 1, we have

$$\begin{pmatrix} b_{11}^1 \\ b_{21}^1 \end{pmatrix} = \frac{\begin{pmatrix} g_{11} & g_{12} \\ g_{21} & g_{22} \end{pmatrix}^{-1} \begin{pmatrix} (\widehat{h}_{21}^1)^* \\ -(\widehat{h}_{11}^1)^* \end{pmatrix}}{\left| \begin{pmatrix} g_{11} & g_{12} \\ g_{21} & g_{22} \end{pmatrix}^{-1} \begin{pmatrix} (\widehat{h}_{21}^1)^* \\ -(\widehat{h}_{11}^1)^* \end{pmatrix} \right|_F}. \tag{4.56}$$

So we know that in order to maximize $|\cos\theta_{hg}^1|$, we can choose $\begin{pmatrix} b_{11}^1 \\ b_{21}^1 \end{pmatrix}$ as described by (4.56) if we have perfect feedback. Since we only have quantized feedback, we should design a codebook in which we can find a vector as close to the one described by (4.56) as possible. So, first, we need to determine the distribution of the optimal $\begin{pmatrix} b_{11}^1 \\ b_{21}^1 \end{pmatrix}$ in (4.56). Note that Eq. (4.56) can also be written as

$$
\begin{pmatrix} b_{11}^1 \\ b_{21}^1 \end{pmatrix} = \eta' \begin{pmatrix} g_{22} & -g_{12} \\ -g_{21} & g_{11} \end{pmatrix} \begin{pmatrix} (\widehat{h}_{21}^1)^* \\ -(\widehat{h}_{11}^1)^* \end{pmatrix}
$$

$$
= \eta' \begin{pmatrix} g_{22}(\widehat{h}_{21}^1)^* + g_{12}(\widehat{h}_{11}^1)^* \\ -g_{21}(\widehat{h}_{21}^1)^* - g_{11}(\widehat{h}_{11}^1)^* \end{pmatrix}
$$

$$
= \eta' \begin{pmatrix} \alpha_1 \\ \alpha_2 \end{pmatrix} \tag{4.57}
$$

where $\eta' = \left| \begin{pmatrix} g_{11} & g_{12} \\ g_{21} & g_{22} \end{pmatrix}^{-1} \begin{pmatrix} (\widehat{h}_{21}^1)^* \\ -(\widehat{h}_{11}^1)^* \end{pmatrix} \right|_F^{-1} |g_{11}g_{22} - g_{21}g_{12}|^{-1}$. Let us assume that the singular value decomposition of $\begin{pmatrix} \alpha_1 \\ \alpha_2 \end{pmatrix}$ is

$$
\begin{pmatrix} \alpha_1 \\ \alpha_2 \end{pmatrix} = \mathbf{U}_\alpha \Sigma_\alpha \mathbf{V}_\alpha^\dagger = \mathbf{U}_\alpha \begin{pmatrix} \lambda_1^\alpha \\ 0 \end{pmatrix} \cdot 1 = \lambda_1^\alpha \cdot \mathbf{U}_\alpha(1). \tag{4.58}
$$

Since $\widehat{h}_{11}^1$ and $\widehat{h}_{21}^1$ are independent from $\mathbf{G}$, conditioned on $\widehat{h}_{11}^1$ and $\widehat{h}_{21}^1$, elements of $\begin{pmatrix} \alpha_1 \\ \alpha_2 \end{pmatrix}$ are all Guassian distributed random variables with the same mean and variance, so any column of $\mathbf{U}_\alpha$ and thus $\begin{pmatrix} \alpha_1 \\ \alpha_2 \end{pmatrix}$ will be an isotropically distributed unitary vector [6]. Further, we can conclude that $\begin{pmatrix} \alpha_1 \\ \alpha_2 \end{pmatrix}$ and thus $\begin{pmatrix} b_{11}^1 \\ b_{21}^1 \end{pmatrix}$ are all isotropically distributed unitary vectors.

Therefore, in order to maximize $|\cos\theta_{hg}^1|$, the codebook for User 2 should provide the best approximation to any isotropically distributed unitary vector and the problem becomes exactly the same as the one we discussed before, i.e., to pack one-dimensional subspaces of a complex space known as Grassmannian line packing. Therefore, the resulting codebook for User 2 will be the same as the codebook Υ_1 for User 1 at time slot 1.

So far, we have shown that by using our codebook, we can maximize $|\mathbf{v}_h^1|$ and $|\cos\theta_{hg}^1|$ at the same time. From (4.43), it is easy to see that the coding gain is maximized.

Similarly, we can prove that in time slot 2, both User 1 and User 2 should adopt the above codebook.

4.5 Comparison of Our Scheme with Two Existing Schemes

In this section, we compare our scheme with two other schemes proposed in the literature. The first scheme is the interference cancellation scheme without feedback proposed in [1, 11]. With the same system model, this scheme can provide a diversity of 2. The second scheme is the interference cancellation scheme with perfect feedback proposed in [12]. With the same system model, this scheme can provide a diversity of 4, i.e., full diversity. We show that our scheme can also provide a diversity of 2 with no feedback. With perfect feedback, our scheme provides the performance of the scheme in [12].

First, let us consider the case without feedback. When the number of feedback bits $K = 0$, we can not choose the best precoders according to the feedback. So our precoders are fixed: in time slot 1, both users use precoder $\begin{pmatrix} 1 \\ 0 \end{pmatrix}$ and in time slot 2, both users use precoder $\begin{pmatrix} 0 \\ 1 \end{pmatrix}$.

By (4.40), we know

$$
\begin{aligned}
P(\mathbf{d} \to \bar{\mathbf{d}}) &\leq E\left[\exp\left(-\frac{1}{4}\rho \left(\left| \frac{(\bar{\mathbf{v}}_g^1)^T}{|\bar{\mathbf{v}}_g^1|}\mathbf{v}_h^1 \right|^2 |\gamma_1 + \gamma_2|^2 \right. \right. \right. \\
&\qquad \left. \left. \left. + \left| \frac{(\bar{\mathbf{v}}_g^2)^T}{|\bar{\mathbf{v}}_g^2|}\mathbf{v}_h^2 \right|^2 |\gamma_1 - \gamma_2|^2 \right) \right) \right] \\
&= E\left[E\left[\exp\left(-\frac{1}{4}\rho \left(\left| \frac{(\bar{\mathbf{v}}_g^1)^T}{|\bar{\mathbf{v}}_g^1|}\mathbf{v}_h^1 \right|^2 |\gamma_1 + \gamma_2|^2 \right. \right. \right. \right. \\
&\qquad \left. \left. \left. \left. + \left| \frac{(\bar{\mathbf{v}}_g^2)^T}{|\bar{\mathbf{v}}_g^2|}\mathbf{v}_h^2 \right|^2 |\gamma_1 - \gamma_2|^2 \right) \right) \middle| \bar{\mathbf{v}}_g^1, \bar{\mathbf{v}}_g^2 \right] \right].
\end{aligned}
\tag{4.59}
$$

Since $\mathbf{v}_h^1 = \begin{pmatrix} h_{11} \\ h_{21} \end{pmatrix}$, $\mathbf{v}_h^2 = \begin{pmatrix} h_{12} \\ h_{22} \end{pmatrix}$, if conditioned on $\bar{\mathbf{v}}_g^1, \bar{\mathbf{v}}_g^2$, both $\frac{(\bar{\mathbf{v}}_g^1)^T}{|\bar{\mathbf{v}}_g^1|}\mathbf{v}_h^1$ and $\frac{(\bar{\mathbf{v}}_g^2)^T}{|\bar{\mathbf{v}}_g^2|}\mathbf{v}_h^2$ are linear combination of independent Gaussian random variables with mean 0 and variance 1. In addition, if conditioned on $\bar{\mathbf{v}}_g^1, \bar{\mathbf{v}}_g^2$, then $\frac{(\bar{\mathbf{v}}_g^1)^T}{|\bar{\mathbf{v}}_g^1|}\mathbf{v}_h^1$ and $\frac{(\bar{\mathbf{v}}_g^2)^T}{|\bar{\mathbf{v}}_g^2|}\mathbf{v}_h^2$ are independent. So we have

$$
P(\mathbf{d} \to \bar{\mathbf{d}}) \leq \frac{1}{(1 + \rho|\gamma_1 + \gamma_2|^2/8)(1 + \rho|\gamma_1 - \gamma_2|^2/8)}.
\tag{4.60}
$$

At high SNRs, one can neglect the one in the denominator and get

$$P(\mathbf{d} \to \bar{\mathbf{d}}) \le \frac{\left(\frac{\rho}{8}\right)^{-2}}{|\gamma_1 + \gamma_2|^2 |\gamma_1 - \gamma_2|^2}. \tag{4.61}$$

It is easy to see that the achievable diversity is 2, which is exactly the same as that of the scheme proposed in [1].

Now we consider the case with perfect feedback. Since the diversity for any $K > 0$ is always 4, obviously, in the case of $K = \infty$, perfect feedback, the diversity of our scheme is the same as that of the scheme proposed in [12].

When there are K bits of feedback, the performance of our system is given by (4.50). We know that as long as the number of feedback bits $K > 0$, our scheme can provide full diversity. Also with the increase of K, the interference term $E\left[\frac{1}{|\cos\theta_{hg}^1|^8}\right]$ decreases to 1. Therefore, the coding gain and the performance of our scheme will approach those of the system with perfect feedback.

4.6 Extension to Any Number of Antennas

In this section, we show that our scheme can also be extended for 2 users with any number of antennas and one receiver with any number of antennas. We will consider two cases. The first one is the case in which the number of transmit antennas N is greater than or equal to the number of receive antennas M. The second one is the case in which $M > N$.

First, we assume $N \ge M$. Similar to the case in Sect. 4.2, User 1 and User 2 transmit Alamouti codes $\mathbf{C}$ and $\mathbf{S}$, respectively. The channels for Users 1 and 2 are

$$\mathbf{H} = [h_{ij}]_{M \times N}, \quad \mathbf{G} = [g_{ij}]_{M \times N}. \tag{4.62}$$

The precoders for Users 1 and 2 are

$$\mathbf{A}^t = [a_{ij}^t]_{N \times 2}, \quad \mathbf{B}^t = [b_{ij}^t]_{N \times 2}. \tag{4.63}$$

Then we can use exactly the same method to design the codebook and precoders. However, when $N \ge M > 2$, with K bits of feedback, the diversity is $M \cdot \min(N, L)$, where $L = 2^K$ is the number of vectors in the codebook. To prove this, we note that in the case of $N \ge M > 2$, (4.44) becomes

$$\begin{aligned}
|\mathbf{v}_h^1|^2 &\ge \frac{|\mathbf{H}\Upsilon_1|^2}{L} = \frac{|\mathbf{H}\mathbf{U}_{\Upsilon_1}\Sigma_{\Upsilon_1}\mathbf{V}_{\Upsilon_1}^{\dagger}|^2}{L} \\
&= \frac{\sum_{j=1}^{L'}(|\lambda_j^{\Upsilon_1}|^2 \sum_{i=1}^{M} |h_{ij}'|^2)}{L}
\end{aligned} \tag{4.64}$$

where $L' = \min(N, L)$. It is easy to see that the number of Gaussian random variables on the right side of (4.64) is ML'. Therefore, when $N \ge M > 2$, following the proof presented in Sect. 4.3, the diversity of our scheme is $M \cdot \min(N, L)$ or $M \cdot \min(N, 2^K)$. In order to achieve a diversity of MN, we need

1. $L \geq N$, i.e., $K \geq \log_2 N$.
2. The rank of matrix $\overline{\Upsilon}_1$ to be N.

Now we consider the case that $N < M$. In this case, we assume the channel matrices and precoders for Users 1 and 2 are given by (4.62) and (4.63). We can use the same method as discussed before to maximize $|v_h^1|$. However, if we want to maximize $|\cos \theta_{hg}^1|^2$, like (4.54), we need to design precoders to make

$$\widetilde{\mathbf{H}}_{M \times 1} = \eta \cdot \mathbf{G}_{M \times N} \cdot \mathbf{B}^1(1)_{N \times 1} \tag{4.65}$$

which means the equivalent signal vectors of the two users are orthogonal to each other. In the above equation, we need to determine N unknown parameters by M equations. Since $N < M$, the number of equations is greater than the number of unknown parameters. Therefore, even with prefect feedback, we cannot find these unknown parameters to satisfy the equations. In other words, since we do not have enough dimensions for precoders, we cannot make $\mathbf{v}_g^i$ orthogonal to $\mathbf{v}_h^i$.

In order to make our proposed scheme extendable to the case of $M > N$, we can choose N receive antennas among all M receive antennas as follows:

In time slot 1, we can choose the N receive antennas such that $\|\mathbf{H}_{\text{new}}\|_F$ is maximized, where $\mathbf{H}_{\text{new}}$ is the new channel matrix with N transmit antennas and the selected N receive antennas. Once the number of receive antennas is equal to the number of transmit antennas, the same method used in Sect. 4.3 can be used to determine the codebook and precoders for Users 1 and 2. At time slot 2, we choose the N receive antennas such that $\|\mathbf{G}_{\text{new}}\|_F$ is maximized, where $\mathbf{G}_{\text{new}}$ is the new channel matrix with N transmit antennas and the selected N receive antennas. Then we design the codebook and precoders for Users 1 and 2 using the same method in the case that $M = N$.

In order to show that we can achieve full diversity for each user using the above proposed method, we consider (4.44). By (4.44), we know

$$
\begin{aligned}
|\mathbf{v}_h^1|^2 &\geq \frac{|\mathbf{H}\Upsilon_1|^2}{L} = \frac{|\mathbf{H}\mathbf{U}_{\Upsilon_1}\Sigma_{\Upsilon_1}\mathbf{V}_{\Upsilon_1}^\dagger|^2}{L} \\
&= \frac{\sum_{j=1}^{N}(|\lambda_j^{\Upsilon_1}|^2 \sum_{i=1}^{N}|h_{ij}'|^2)}{L} \\
&\geq \frac{|\lambda_{\min}^{\Upsilon_1}|^2 \sum_{j=1}^{N}\sum_{i=1}^{N}|h_{ij}'|^2}{L} \\
&= \frac{|\lambda_{\min}^{\Upsilon_1}|^2 \sum_{j=1}^{N}\sum_{i=1}^{N}|h_{ij}|^2}{L} \\
&= \frac{|\lambda_{\min}^{\Upsilon_1}|^2 \|\mathbf{H}_{\text{new}}\|_F^2}{L}.
\end{aligned}
\tag{4.66}
$$

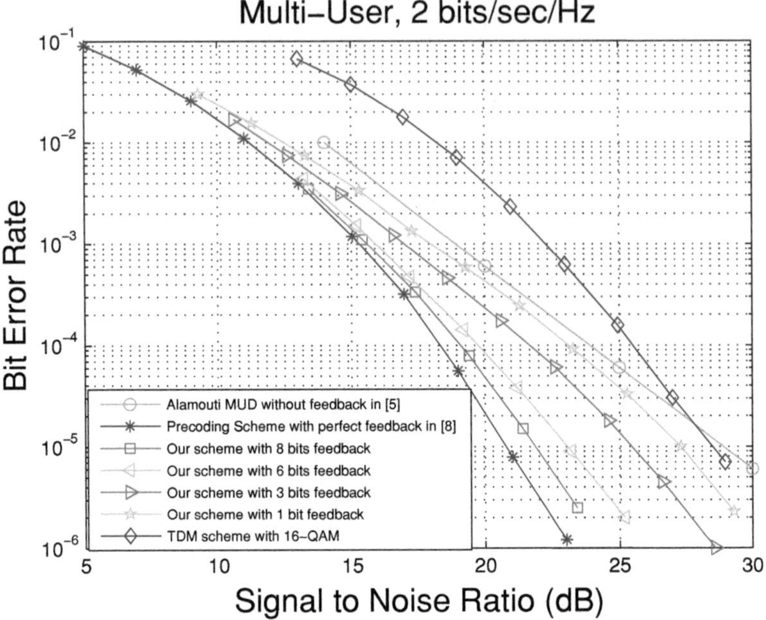

Fig. 4.3 Comparison of our scheme, Alamouti MUD in [1] and Precoding scheme in [13] for 2 users each with 2 transmit antennas and 1 receiver with 2 receive antennas

Since we know $||\mathbf{H}_{\text{new}}||_F^2$ is maximized, the average of the norms of all columns in matrix $\mathbf{H}_{\text{new}}$ will be no less than the average of the norms of all columns in matrix $\mathbf{H}$, i.e.,

$$\frac{||\mathbf{H}_{\text{new}}||_F^2}{N} = \frac{\sum_{i=1}^{N} |\mathbf{H}_{\text{new}}(i)|^2}{N} \geq \frac{||\mathbf{H}||_F^2}{M} = \frac{\sum_{i=1}^{M} |\mathbf{H}(i)|^2}{M}. \tag{4.67}$$

Substituting (4.67) to (4.66), we have

$$|\mathbf{v}_h^1|^2 \geq \frac{|\lambda_{\min}^{\Upsilon_1}|^2 ||\mathbf{H}_{\text{new}}||_F^2}{L} \geq \frac{N|\lambda_{\min}^{\Upsilon_1}|^2 ||\mathbf{H}||_F^2}{ML}$$
$$= \frac{N|\lambda_{\min}^{\Upsilon_1}|^2 \sum_{j=1}^{N} \sum_{i=1}^{M} |h_{ij}|^2}{ML}. \tag{4.68}$$

Since there are MN Gaussian random variables on the right side of (4.68), it is easy to prove that User 1 can achieve a diversity of MN, i.e., full diversity. Similarly, it can be proved that User 2 can also achieve full diversity. When there are more than two users, there will be more interference to be dealt with. The precoding and decoding scheme will be more complex. Due to the limitation of the space, we leave the extension of the scheme to more than two users as our future work.

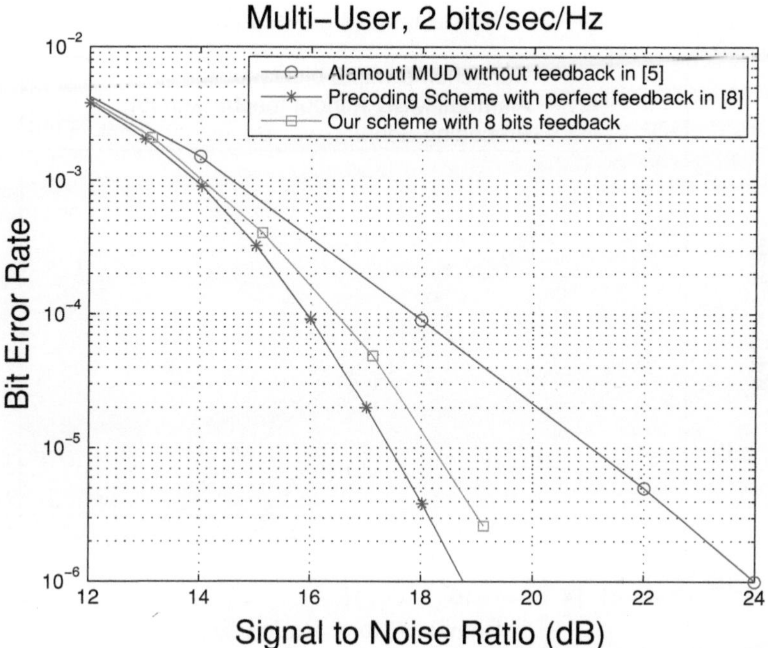

Fig. 4.4 Comparison of our scheme, Alamouti MUD in [1] and Precoding scheme in [13] for 2 users each with 4 transmit antennas and 1 receiver with 2 receive antennas

4.7 Simulation Results

In this section, we provide simulation results that confirm our analysis in the previous sections. We assume a quasi-static Rayleigh fading channel. The performance of our proposed scheme is shown in Figs. 4.3, 4.4 and 4.5. In each figure, the curves for Users 1 and 2 are identical. In Fig. 4.3, we consider 2 users each equipped with 2 transmit antennas and a receiver with 2 receive antennas. We compare our results using QPSK with the results in [1] for the same configuration without channel information at the transmitter and the results in [12] for the same configuration with perfect feedback. Note that if the feedback is zero in our system (no channel information), we can pick an identity matrix as our precoder and our transmitter will be the same as the transmitter in [1]. In fact, this backward compatibility is the main reason for using an Alamouti code. Otherwise, our scheme also works for other full rate space time codes and all the above derivations are still valid.

In order to illustrate the effect of the number of bits, we provide the performance with 1, 3, 6, 8 bits feedback, respectively. It can be seen that with 2 receive antennas, the multi-user detection (MUD) method proposed in [1] can cancel the interference but only provides a diversity of 2. The scheme proposed in [12] with perfect feedback can achieve interference cancellation and provide a diversity of 4, full diversity. In comparison, using the proposed scheme in this chapter, we can also achieve

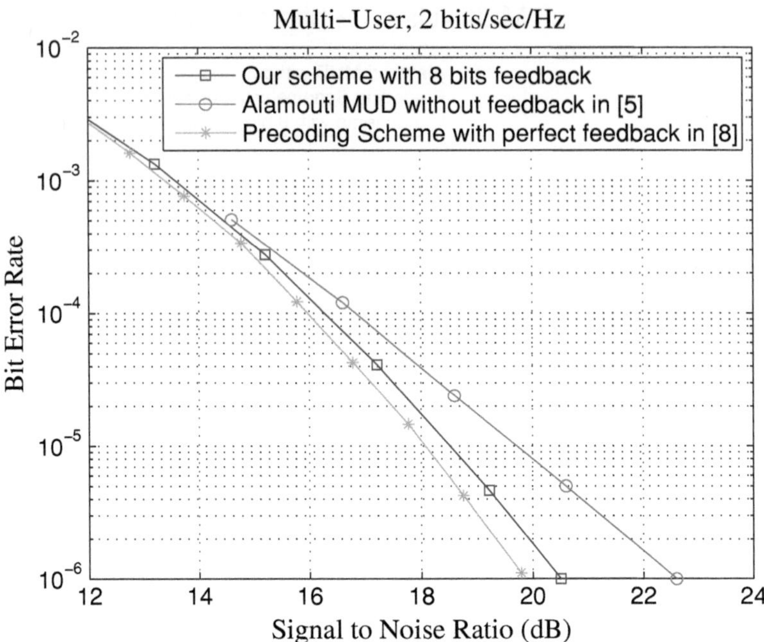

Fig. 4.5 Comparison of our scheme, Alamouti MUD in [1] and Precoding scheme in [13] for 2 users each with 2 transmit antennas and 1 receiver with 3 receive antennas

interference cancellation as well as full diversity only with quantized feedback, even with only 1 bit of feedback. But the performance highly depends on the number of feedback bits. When the number of feedback bits is small, the performance of our scheme is close to the performance of the scheme without feedback. When the number of feedback bits increases, the performance will approach the performance of the system with perfect feedback. Therefore, our proposed scheme provides a solution to fill the performance gap between [1] and [12]. Finally, we also provide the simulation results for the time-division multiplexing (TDM) case in which the two users transmit Alamouti codes in different time slots. In this case, there will be no interference at all. In order to match the rate, each user adopts 16-QAM. From the simulation results, we can see that although the TDM scheme can achieve full diversity and the decoding complexity is low, it will lose coding gain.

In Fig. 4.4, we provide the performance of our scheme with 8 bits of feedback for 2 users each with 4 transmit antennas and one receiver with 2 receive antennas. Also we compare the performance of our scheme with the schemes in [1] and [12]. It is easy to see that our scheme with 8 bits of feedback has achieved full diversity and has outperformed the scheme in [1]. Compared with the scheme with perfect feedback, the performance difference is about 1 dB.

In Fig. 4.5, we present the performance of our scheme with 8 bits of feedback for 2 users each with 2 transmit antennas and one receiver with 3 receive antennas. Once again, the performance of our scheme outperforms the performance of the scheme in [1] and approaches the performance of the scheme in [12]. Simulation results show that by using only a few bits of feedback, one can approach the performance of a system with perfect feedback.

4.8 Conclusions

In this chapter, we investigate how to cancel the interference and achieve full diversity for two users with two transmit antennas and one receiver with two receive antennas in a multiple access channel using quantized feedback. Using quantized feedback, we propose the precoding and decoding method, the feedback scheme and the codebook design to cancel interference and achieve full diversity. Also we show that the performance of our proposed scheme is determined by the number of feedback bits. With the increase of the feedback bits, the performance of our scheme approaches that of the system with perfect feedback. Finally we extend our scheme to two users with any number of transmit antennas and one receiver with any number of receive antennas. Simulation results are provided to confirm our analytical results.

References

1. Kazemitabar, J., Jafarkhani, H.: Multiuser interference cancellation and detection for users with more than two transmit antennas. IEEE Trans. Commun. **56**(4), 574–583 (2008)
2. Ghaderipoor, A., Tellambura, C.: Optimal precoder for rate less than one space-time block codes. In: Proceedings of IEEE International Conference on Communications, Glasgow, June (2007)
3. Davis, P.J.: Circulant Matrices. Wiley, New York (1979)
4. Simon, M.K., Alouini, M.-S.: Digital Communication over Fading Channels. Wiley, New York (2000)
5. Goldsmith, A.J., Jafar, S.A., Jindal, N., Vishwanath, S.: Capacity limits of MIMO channels. IEEE J. Sel. Areas Commun. **21**(5), 684–702 (2003)
6. Marzetta, T.L., Hochwald, B.M.: Capacity of a mobile multiple-antenna communication link in Rayleigh flat fading. IEEE Trans. Inf. Theory **45**, 139–157 (1999)
7. Conway, J.H., Hardin, R.H., Sloane, N.J.A.: Packing lines, planes, etc.: packings in Grassmannian spaces. Exp. Math. **5**(2), 139–159 (1996)
8. Strohmer, T., Heath Jr, R.W.: Grassmannian frames with applications to coding and communications. Appl. Comput. Harmon. Anal. **14**(3), 257–275 (2003)
9. Love, D.J., Heath Jr. R.W., Strohmer, T.: Grassmannian beamforming for multiple-input multiple-output wireless systems. IEEE Trans. Info. Theory **49**(10), 2735–2747 (2003).
10. Agrawal, D., Richardson, T.J., Urbanke, R.L.: Multiple-antenna signal constellations for fading channels. IEEE Trans. Inf. Theory **47**, 2618–2626 (2001)
11. Kazemitabar, J., Jafarkhani, H., Performance analysis of multiple-antenna multi-user detection. In: Proceedings of, : Workshop on Information Theory and its Applications. San Diego, February (2009). 2009.
12. Li, F., Jafarkhani, H.: Interference cancellation and detection using precoders. In: Proceedings of IEEE International Conference on Communications, Dresden, June (2009)
13. Li, F., Jafarkhani, H.: Multiple-antenna interference cancellation and detection for two users using precoders. IEEE J. Sel. Topics Signal Process. **3**(6), 1066–1078 (2009)

Chapter 5
Interference-Free Transmission for X channels

5.1 Channel Model

We introduce our channel model as shown in Fig. 5.1. We assume there are 2 users each with N transmit antennas and 2 receivers each with M receive antennas. Both users want to send different codewords to Receivers 1 and 2 on the same frequency band at the same time. As shown in Fig. 5.1, User 1 wants to send codeword C_1 to Receiver 1 and C_2 to Receiver 2. User 2 wants to send codeword S_1 to Receiver 1 and S_2 to Receiver 2. We also assume that full channel information is available at each user and receiver. The problem we want to solve is how to derive interference-free codewords from each user at each receiver with full diversity and rate 1. We let each user transmit Quasi Orthogonal Space-Time Block Codes (QOSTBCs) [1] as follows:

$$\mathbf{C}_i = \begin{pmatrix} c_{i1} & -c_{i2}^* & c_{i3} & -c_{i4}^* \\ c_{i2} & c_{i1}^* & c_{i4} & c_{i3}^* \\ c_{i3} & -c_{i4}^* & c_{i1} & -c_{i2}^* \\ c_{i4} & c_{i3}^* & c_{i2} & c_{i1}^* \end{pmatrix}, \quad \mathbf{S}_j = \begin{pmatrix} s_{j1} & -s_{j2}^* & s_{j3} & -s_{j4}^* \\ s_{j2} & s_{j1}^* & s_{j4} & s_{j3}^* \\ s_{j3} & -s_{j4}^* & s_{j1} & -s_{j2}^* \\ s_{j4} & s_{j3}^* & s_{j2} & s_{j1}^* \end{pmatrix} \tag{5.1}$$

where $i, j = 1, 2$. Note that we can also use other space-time codes with rate one and QOSTBC is just one example. Since User 1 needs to send $\mathbf{C}_1$ to Receiver 1 and $\mathbf{C}_2$ to Receiver 2 at the same time, we can let User 1 transmit the following combined codewords at time slot t

$$\mathbf{C}^t = \mathbf{A}_1^t \mathbf{C}_1(t) + \mathbf{A}_2^t \mathbf{C}_2(t) \tag{5.2}$$

where

$$\mathbf{A}_l^t = [a_1^t(i, j)]_{N \times 4}, \quad t = 1, 2, 3, 4, \quad l = 1, 2 \tag{5.3}$$

are the precoders we need to design for User 1. Note that in order to satisfy the power constraint, we need

$$\|\mathbf{A}_1^t\|_F^2 + \|\mathbf{A}_2^t\|_F^2 = 1 \tag{5.4}$$

F. Li, *Interference Cancellation Using Space-Time Processing and Precoding Design*, Signals and Communication Technology, DOI: 10.1007/978-3-642-30712-6_5,

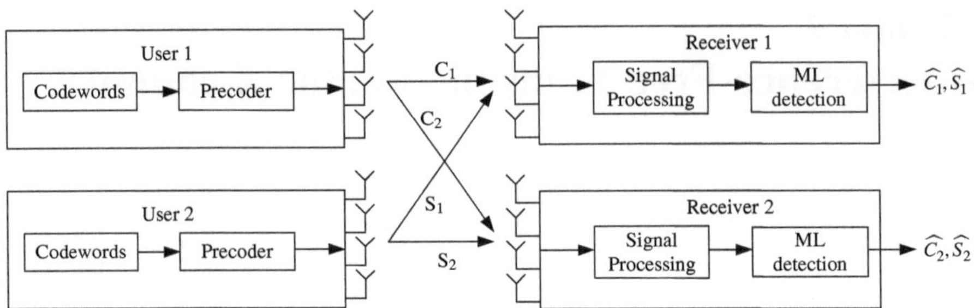

Fig. 5.1 X channel

In this chapter, we assume that $||\mathbf{A}_1^t||_F^2=||\mathbf{A}_2^t||_F^2=\frac{1}{2}$. Similarly, User 2 sends the following codewords

$$\mathbf{S}^t = \mathbf{B}_1^t \mathbf{S}_1(t) + \mathbf{B}_2^t \mathbf{S}_2(t) \tag{5.5}$$

with the power constraint

$$||\mathbf{B}_1^t||_F^2 + ||\mathbf{B}_2^t||_F^2 = 1 \tag{5.6}$$

where

$$\mathbf{B}_l^t = [b_l^t(i, j)]_{N \times 4}, \quad t = 1, 2, 3, 4, \quad l = 1, 2 \tag{5.7}$$

are the precoders we need to design for User 2. Also we assume that $||\mathbf{B}_1^t||_F^2 = ||\mathbf{B}_2^t||_F^2 = \frac{1}{2}$. The channels are quasi-static flat Rayleigh fading and keep unchanged during four time slots. Then we let

$$\mathbf{H}_l = [h_l(i, j)]_{M \times N}, \quad l = 1, 2 \tag{5.8}$$

denote the channel matrix between User 1 and Receivers l, respectively. Similarly, we use

$$\mathbf{G}_l = [g_l(i, j)]_{M \times N}, \quad l = 1, 2 \tag{5.9}$$

to denote the channel matrix between User 2 and Receiver l, respectively. Then the received signals at Receiver 1 at time slot t can be denoted by

$$\mathbf{y}_1^t = \mathbf{H}_1 \mathbf{A}_1^t \mathbf{C}_1(t) + \mathbf{H}_1 \mathbf{A}_2^t \mathbf{C}_2(t) + \mathbf{G}_1 \mathbf{B}_1^t \mathbf{S}_1(t) + \mathbf{G}_1 \mathbf{B}_2^t \mathbf{S}_2(t) + \mathbf{n}_1^t \tag{5.10}$$

where

$$\mathbf{y}_1^t = [y_1^t(i, 1)]_{M \times 1}, \quad \mathbf{n}_1^t = [n_1^t(i, 1)]_{M \times 1} \tag{5.11}$$

denote the received signals and the noise at Receiver 1, respectively, at time slot t. Similarly, at time slot t, Receiver 2 will receive the following signals

$$\mathbf{y}_2^t = \mathbf{H}_2 \mathbf{A}_1^t \mathbf{C}_1(t) + \mathbf{H}_2 \mathbf{A}_2^t \mathbf{C}_2(t) + \mathbf{G}_2 \mathbf{B}_1^t \mathbf{S}_1(t) + \mathbf{G}_2 \mathbf{B}_2^t \mathbf{S}_2(t) + \mathbf{n}_2^t \tag{5.12}$$

where

$$\mathbf{y}_2^t = [y_2^t(i, 1)]_{M \times 1}, \quad \mathbf{n}_2^t = [n_2^t(i, 1)]_{M \times 1} \tag{5.13}$$

Equations (5.10) and (5.12) are the channel equations on which we will base our design in this chapter.

5.2 Precoder Design

In this chapter, we aim to design precoders to achieve the following two goals:

1. At each time slot, each receiver can obtain interference-free signals from each user when all the users transmit at the same time.
2. Our system can provide full diversity for each user.

The first goal is easy to understand. The second goal needs explanation as different users and different codewords may have different diversities. Full diversity for User 1 means at Receiver 1, the diversity for codeword $\mathbf{C}_1$ is full and at Receiver 2, the diversity for codeword $\mathbf{C}_2$ is full. Similarly, by saying the diversity for User 2 is full, we mean that at Receiver 1, the diversity for codeword $\mathbf{S}_1$ is full and at Receiver 2, the diversity for codeword $\mathbf{S}_2$ is full. In this section, we show our main idea to achieve interference-free transmission. Later, we will show that based on our proposed interference-free transmission scheme in this section, we can further achieve full diversity.

Our main idea to achieve interference-free transmission is to adjust each signal in the signal space of X channels by using precoders for each transmitter, such that at the receiver each desired signal is orthogonal to all other signals. In this way, we can achieve interference-free transmission. To make our scheme easier to understand, we will start our design for the case with $M = 4$ first and see the minimum number of transmit antennas needed to achieve interference-free transmission. Later we will generalize our scheme for any N and M.

In Eq. (5.10), we use

$$\mathbf{H}_{11}^t = \mathbf{H}_1\mathbf{A}_1^t, \mathbf{H}_{12}^t = \mathbf{H}_1\mathbf{A}_2^t, \mathbf{G}_{11}^t = \mathbf{G}_1\mathbf{B}_1^t, \mathbf{G}_{12}^t = \mathbf{G}_1\mathbf{B}_2^t \tag{5.14}$$

to denote the equivalent channel matrices. Then Eq. (5.10) becomes

$$\mathbf{y}_1^t = \mathbf{H}_{11}^t\mathbf{C}_1(t) + \mathbf{H}_{12}^t\mathbf{C}_2(t) + \mathbf{G}_{11}^t\mathbf{S}_1(t) + \mathbf{G}_{12}^t\mathbf{S}_2(t) + \mathbf{n}_1^t \tag{5.15}$$

Similarly, in Eq. (5.12), if we use

$$\mathbf{H}_{21}^t = \mathbf{H}_2\mathbf{A}_1^t, \mathbf{H}_{22}^t = \mathbf{H}_2\mathbf{A}_2^t, \mathbf{G}_{21}^t = \mathbf{G}_2\mathbf{B}_1^t, \mathbf{G}_{22}^t = \mathbf{G}_2\mathbf{B}_2^t \tag{5.16}$$

to denote the equivalent channel matrices, we have

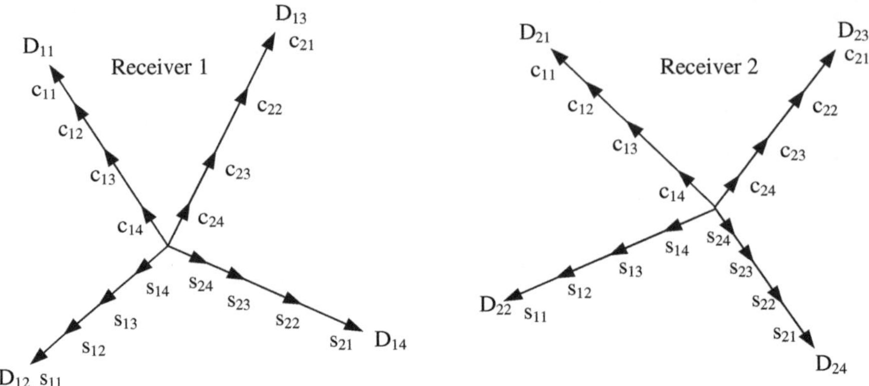

Fig. 5.2 Signal vector illustration at two receivers

$$\mathbf{y}_2^t = \mathbf{H}_{21}^t \mathbf{C}_1(t) + \mathbf{H}_{22}^t \mathbf{C}_2(t) + \mathbf{G}_{21}^t \mathbf{S}_1(t) + \mathbf{G}_{22}^t \mathbf{S}_2(t) + \mathbf{n}_2^t \qquad (5.17)$$

By Eq. (5.15), since the receiver has four receive antennas, each symbol is actually transmitted along a 4-dimensional vector in a 4-dimensional space. Because each user sends eight symbols at the same time, at the receiver, there are 16 signal vectors in the four-dimensional space.

Since we want to send $\mathbf{C}_1$, $\mathbf{C}_2$, $\mathbf{S}_1$, $\mathbf{S}_2$ without any interference from each other, we let each one of $\mathbf{C}_1$, $\mathbf{C}_2$, $\mathbf{S}_1$, $\mathbf{S}_2$ occupy only one dimension. In other words, for any codeword, we should transmit each of the corresponding four symbols in the same direction. In this way, there are only four transmit directions. Once we can align the four transmit directions of $\mathbf{C}_1$, $\mathbf{C}_2$, $\mathbf{S}_1$, $\mathbf{S}_2$ properly, we can separate them completely. This is our first step to achieve interference-free transmission.

This idea is illustrated in Fig. 5.2, where $\mathbf{D}_{ij}$ is the jth direction at Receiver i. By Eq. (5.15), c_{11}, c_{12}, c_{13}, c_{14} are transmitted along $\mathbf{H}_{11}^t(1)$, $\mathbf{H}_{11}^t(2)$, $\mathbf{H}_{11}^t(3)$, $\mathbf{H}_{11}^t(4)$, respectively. In order to make $\mathbf{H}_{11}^t(1)$, $\mathbf{H}_{11}^t(2)$, $\mathbf{H}_{11}^t(3)$, $\mathbf{H}_{11}^t(4)$ along the same direction, by Eq. (5.14), we need

$$\mathbf{A}_1^t(1) = \frac{1}{\alpha_{11}^t} \mathbf{A}_1^t(2) = \frac{1}{\alpha_{12}^t} \mathbf{A}_1^t(3) = \frac{1}{\alpha_{13}^t} \mathbf{A}_1^t(4) \qquad (5.18)$$

where α_{11}^t, α_{12}^t, α_{13}^t are constants that we will determine later. From $||\mathbf{A}_1^t||_F^2 = \frac{1}{2}$, we know

$$||\mathbf{A}_1^t(1)||_F^2 = \frac{1}{2(1 + (\alpha_{11}^t)^2 + (\alpha_{12}^t)^2 + (\alpha_{13}^t)^2)} \qquad (5.19)$$

So when we design precoder $\mathbf{A}_1^t$, Eqs. (5.18) and (5.19) should be satisfied. Similarly, precoders $\mathbf{A}_2^t$, $\mathbf{B}_1^t$, $\mathbf{B}_2^t$ should also satisfy the following conditions:

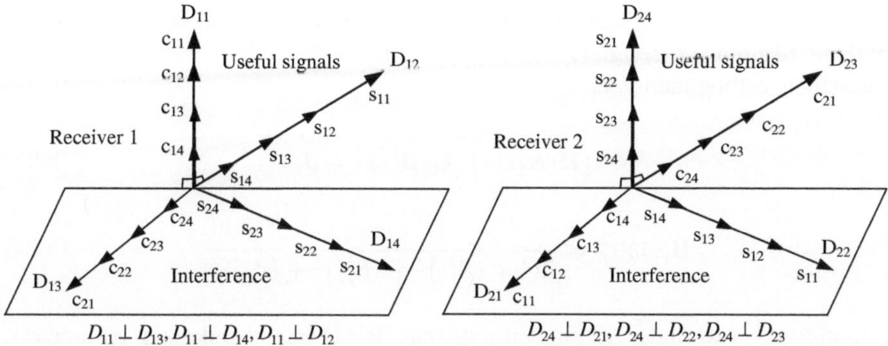

Fig. 5.3 Signal vector illustration for our interference cancellation scheme

$$\mathbf{A}_2^t(1) = \frac{1}{\alpha_{21}^t}\mathbf{A}_2^t(2) = \frac{1}{\alpha_{22}^t}\mathbf{A}_2^t(3) = \frac{1}{\alpha_{23}^t}\mathbf{A}_2^t(4) \tag{5.20}$$

$$\mathbf{B}_1^t(1) = \frac{1}{\beta_{11}^t}\mathbf{B}_1^t(2) = \frac{1}{\beta_{12}^t}\mathbf{B}_1^t(3) = \frac{1}{\beta_{13}^t}\mathbf{B}_1^t(4) \tag{5.21}$$

$$\mathbf{B}_2^t(1) = \frac{1}{\beta_{21}^t}\mathbf{B}_2^t(2) = \frac{1}{\beta_{22}^t}\mathbf{B}_2^t(3) = \frac{1}{\beta_{23}^t}\mathbf{B}_2^t(4) \tag{5.22}$$

Next, we will discuss how to design precoders to align the four directions at each receiver to achieve interference cancellation. The main idea is to make each signal vector of useful signals orthogonal to all other signal vectors. Or in other words, all the interference lies in a subspace which is orthogonal to the useful signals. We call this method Scheme I. For example, for Receiver 1, only $\mathbf{C}_1, \mathbf{S}_1$ are useful codewords and $\mathbf{C}_2, \mathbf{S}_2$ are not decoded. Therefore, we can consider $\mathbf{C}_2, \mathbf{S}_2$ as interference and align them in a subspace that is orthogonal to the two orthogonal vectors for $\mathbf{C}_1, \mathbf{S}_1$. Similarly, for Receiver 2, $\mathbf{C}_1, \mathbf{S}_1$ can be considered as interference. So we align them in a subspace that is orthogonal to the two orthogonal vectors for $\mathbf{C}_2, \mathbf{S}_2$.

This idea is illustrated in Fig. 5.3. At Receiver 1, the useful signal vector $\mathbf{D}_{11}$ is orthogonal to the other useful signal vector $\mathbf{D}_{12}$. Both $\mathbf{D}_{11}$ and $\mathbf{D}_{12}$ are orthogonal to the interference subspace created by the two interference vectors $\mathbf{D}_{13}, \mathbf{D}_{14}$. At Receiver 2, the useful signal vector $\mathbf{D}_{23}$ is orthogonal to the other useful signal vector $\mathbf{D}_{24}$. Both $\mathbf{D}_{23}$ and $\mathbf{D}_{24}$ are orthogonal to the interference subspace created by the two interference vectors created by the two interference vectors $\mathbf{D}_{21}, \mathbf{D}_{22}$. After aligning the signal vectors as shown in Fig. 5.3, it is easy to see that the desired signals at each receiver are free of interference.

In what follows, we will show the above idea is achievable, i.e., we can always find proper precoders to align signal vectors as shown in Fig. 5.3, and the minimum N required. For the sake of simplicity, we only consider the first time slot. The illustration for other time slots is similar. First, we design the precoder $\mathbf{A}_1^1$ for $\mathbf{C}_1$. Since the whole space is available, we can pick a precoder to optimize the transmission of $\mathbf{C}_1$, for example, we can use beamforming. Assume that we have already designed the precoder $\mathbf{A}_1^1$ for $\mathbf{C}_1$. Then the directions $\mathbf{D}_{11}$ at Receiver 1 and $\mathbf{D}_{21}$ at Receiver 2 are fixed. Second, we design $\mathbf{B}_1^1$ for $\mathbf{S}_1$ to make $\mathbf{D}_{12}\perp\mathbf{D}_{11}$, where we use the symbol $\perp$ to denote orthogonality, i.e.,

$$\left(\mathbf{H}_1\mathbf{A}_1^1(1)\right)^{\dagger}\mathbf{G}_1\mathbf{B}_1^1(1) = 0 \tag{5.23}$$

$$\|\mathbf{B}_1^1(1)\|_F^2 = \frac{1}{2(1 + (\beta_{11}^1)^2 + (\beta_{12}^1)^2 + (\beta_{13}^1)^2)} \tag{5.24}$$

Note that when we have N transmit antennas, $\mathbf{B}_1^1(1)$ has N unknown parameters. Therefore, we will have 2 equations and N unknown parameters to solve. Once $N \geq 2$, the solution always exists. Third, we design $\mathbf{A}_2^1$ for $\mathbf{C}_2$ to make $\mathbf{D}_{13}\perp\mathbf{D}_{11}$, $\mathbf{D}_{13}\perp\mathbf{D}_{12}$, $\mathbf{D}_{23}\perp\mathbf{D}_{21}$, $\mathbf{D}_{23}\perp\mathbf{D}_{22}$, i.e.,

$$\left(\mathbf{H}_1\mathbf{A}_1^1(1)\right)^{\dagger}\mathbf{H}_1\mathbf{A}_2^1(1) = 0 \tag{5.25}$$

$$\left(\mathbf{G}_1\mathbf{B}_1^1(1)\right)^{\dagger}\mathbf{H}_1\mathbf{A}_2^1(1) = 0 \tag{5.26}$$

$$\left(\mathbf{H}_2\mathbf{A}_1^1(1)\right)^{\dagger}\mathbf{H}_2\mathbf{A}_2^1(1) = 0 \tag{5.27}$$

$$\left(\mathbf{G}_2\mathbf{B}_1^1(1)\right)^{\dagger}\mathbf{H}_2\mathbf{A}_2^1(1) = 0 \tag{5.28}$$

$$\|\mathbf{A}_2^1(1)\|_F^2 = \frac{1}{2(1 + (\alpha_{21}^1)^2 + (\alpha_{22}^1)^2 + (\alpha_{23}^1)^2)} \tag{5.29}$$

There are 5 equations and N unknown parameters. As long as $N \geq 5$, we can still find a solution. Finally, we design $\mathbf{B}_2^1$ for $\mathbf{S}_2$ to make $\mathbf{D}_{14}\perp\mathbf{D}_{11}$, $\mathbf{D}_{14}\perp\mathbf{D}_{12}$, $\mathbf{D}_{24}\perp\mathbf{D}_{21}$, $\mathbf{D}_{24}\perp\mathbf{D}_{22}$, $\mathbf{D}_{24}\perp\mathbf{D}_{23}$, i.e.,

$$\left(\mathbf{H}_1\mathbf{A}_1^1(1)\right)^{\dagger}\mathbf{G}_1\mathbf{B}_2^1(1) = 0 \tag{5.30}$$

$$\left(\mathbf{G}_1\mathbf{B}_1^1(1)\right)^{\dagger}\mathbf{G}_1\mathbf{B}_2^1(1) = 0 \tag{5.31}$$

$$\left(\mathbf{H}_2\mathbf{A}_1^1(1)\right)^{\dagger}\mathbf{G}_2\mathbf{B}_2^1(1) = 0 \tag{5.32}$$

$$\left(\mathbf{G}_2\mathbf{B}_1^1(1)\right)^{\dagger}\mathbf{G}_2\mathbf{B}_2^1(1) = 0 \tag{5.33}$$

$$\left(\mathbf{H}_2\mathbf{A}_2^1(1)\right)^{\dagger}\mathbf{G}_2\mathbf{B}_2^1(1) = 0 \tag{5.34}$$

$$\|\mathbf{B}_1^1(1)\|_F^2 = \frac{1}{2(1 + (\beta_{11}^1)^2 + (\beta_{12}^1)^2 + (\beta_{13}^1)^2)} \tag{5.35}$$

There are 6 equations and N unknown parameters. As long as $N \geq 6$, we can always find a solution. Therefore, using the new signal vector structure, we can design precoders $\mathbf{A}_1^1$, $\mathbf{A}_2^1$, $\mathbf{B}_1^1$, $\mathbf{B}_2^1$ to achieve interference-free transmission for each codeword at both receivers as shown in Fig. 5.3 when we have at least six transmit antennas.

5.3 Decoding

In the last section, we have shown how to achieve interference-free transmission. Once interference-free transmission as shown in Fig. 5.3 is achieved, it is easy to realize low-complexity decoding. In this section, we will show how to decode and analyze the decoding complexity. Based on Eq. (5.15), after combining the channel equations in the first four time slots and making some simple transformation, we have

$$\tilde{\mathbf{y}}_1 = \tilde{\mathbf{H}}_1 \begin{pmatrix} c_{11} \\ c_{12} \\ c_{13} \\ c_{14} \end{pmatrix} + \tilde{\mathbf{H}}_2 \begin{pmatrix} c_{21} \\ c_{22} \\ c_{23} \\ c_{24} \end{pmatrix} + \tilde{\mathbf{G}}_1 \begin{pmatrix} s_{11} \\ s_{12} \\ s_{13} \\ s_{14} \end{pmatrix} + \tilde{\mathbf{G}}_2 \begin{pmatrix} s_{21} \\ s_{22} \\ s_{23} \\ s_{24} \end{pmatrix} + \tilde{\mathbf{n}}_1 \tag{5.36}$$

where

$$\tilde{\mathbf{H}}_k = \begin{pmatrix} \mathbf{H}_{1k}^1(1) & \mathbf{H}_{1k}^1(2) & \mathbf{H}_{1k}^1(3) & \mathbf{H}_{1k}^1(4) \\ (\mathbf{H}_{1k}^2(2))^* & -(\mathbf{H}_{1k}^2(1))^* & (\mathbf{H}_{1k}^2(4))^* & -(\mathbf{H}_{1k}^2(3))^* \\ \mathbf{H}_{1k}^3(3) & \mathbf{H}_{1k}^3(4) & \mathbf{H}_{1k}^3(1) & \mathbf{H}_{1k}^3(2) \\ (\mathbf{H}_{1k}^4(4))^* & -(\mathbf{H}_{1k}^4(3))^* & (\mathbf{H}_{1k}^4(2))^* & -(\mathbf{H}_{1k}^4(1))^* \end{pmatrix}, \quad \tilde{\mathbf{y}}_1 = \begin{pmatrix} \mathbf{y}_1^1 \\ (\mathbf{y}_1^2)^* \\ \mathbf{y}_1^3 \\ (\mathbf{y}_1^4)^* \end{pmatrix} \tag{5.37}$$

$$\tilde{\mathbf{G}}_k = \begin{pmatrix} \mathbf{G}_{1k}^1(1) & \mathbf{G}_{1k}^1(2) & \mathbf{G}_{1k}^1(3) & \mathbf{G}_{1k}^1(4) \\ (\mathbf{G}_{1k}^2(2))^* & -(\mathbf{G}_{1k}^2(1))^* & (\mathbf{G}_{1k}^2(4))^* & -(\mathbf{G}_{1k}^2(3))^* \\ \mathbf{G}_{1k}^3(3) & \mathbf{G}_{1k}^3(4) & \mathbf{G}_{1k}^3(1) & \mathbf{G}_{1k}^3(2) \\ (\mathbf{G}_{1k}^4(4))^* & -(\mathbf{G}_{1k}^4(3))^* & (\mathbf{G}_{1k}^4(2))^* & -(\mathbf{G}_{1k}^4(1))^* \end{pmatrix}, \quad \tilde{\mathbf{n}}_1 = \begin{pmatrix} \mathbf{n}_1^1 \\ (\mathbf{n}_1^2)^* \\ \mathbf{n}_1^3 \\ (\mathbf{n}_1^4)^* \end{pmatrix} \tag{5.38}$$

where $k = 1, 2$. By the property of our precoders in Eqs. (5.18), (5.20–5.22), $\tilde{\mathbf{H}}_k$ and $\tilde{\mathbf{G}}_k$ can also be written as

$$\tilde{\mathbf{H}}_k = \begin{pmatrix} \mathbf{H}_{1k}^1(1) & \alpha_{11}^1 \mathbf{H}_{1k}^1(1) & \alpha_{12}^1 \mathbf{H}_{1k}^1(1) & \alpha_{13}^1 \mathbf{H}_{1k}^1(1) \\ (\alpha_{11}^2 \mathbf{H}_{1k}^2(1))^* & -(\mathbf{H}_{1k}^2(1))^* & (\alpha_{13}^2 \mathbf{H}_{1k}^2(1))^* & -(\alpha_{12}^2 \mathbf{H}_{1k}^2(1))^* \\ \alpha_{12}^3 \mathbf{H}_{1k}^3(1) & \alpha_{13}^3 \mathbf{H}_{1k}^3(1) & \mathbf{H}_{1k}^3(1) & \alpha_{11}^3 \mathbf{H}_{1k}^3(1) \\ (\alpha_{13}^4 \mathbf{H}_{1k}^4(1))^* & -(\alpha_{12}^4 \mathbf{H}_{1k}^4(1))^* & (\alpha_{11}^4 \mathbf{H}_{1k}^4(1))^* & -(\mathbf{H}_{1k}^4(1))^* \end{pmatrix} \tag{5.39}$$

$$\tilde{\mathbf{G}}_k = \begin{pmatrix} \mathbf{G}^1_{1k}(1) & \beta^1_{11}\mathbf{G}^1_{1k}(1) & \beta^1_{12}\mathbf{G}^1_{1k}(1) & \beta^1_{13}\mathbf{G}^1_{1k}(1) \\ (\beta^2_{11}\mathbf{G}^2_{1k}(1))^* & -(\mathbf{G}^2_{1k}(1))^* & (\beta^2_{13}\mathbf{G}^2_{1k}(1))^* & -(\beta^2_{12}\mathbf{G}^2_{1k}(1))^* \\ \beta^3_{12}\mathbf{G}^3_{1k}(1) & \beta^3_{13}\mathbf{G}^3_{1k}(1) & \mathbf{G}^3_{1k}(1) & \beta^3_{11}\mathbf{G}^3_{1k}(1) \\ (\beta^4_{13}\mathbf{G}^4_{1k}(1))^* & -(\beta^4_{12}\mathbf{G}^4_{1k}(1))^* & (\beta^4_{11}\mathbf{G}^4_{1k}(1))^* & -(\mathbf{G}^4_{1k}(1))^* \end{pmatrix} \tag{5.40}$$

where $k = 1, 2$. In addition, by using our precoders, we have

$$(\mathbf{H}^t_{11}(1))^\dagger \mathbf{H}^t_{12}(1) = (\mathbf{H}^t_{11}(1))^\dagger \mathbf{G}^t_{11}(1) = (\mathbf{H}^t_{11}(1))^\dagger \mathbf{G}^t_{12}(1) = (\mathbf{G}^t_{11}(1))^\dagger \mathbf{H}^t_{12}(1)$$
$$= (\mathbf{G}^t_{11}(1))^\dagger \mathbf{G}^t_{12}(1) = 0 \tag{5.41}$$

where $t = 1, 2, 3, 4$. Equation (5.41) means that at time slots 1, 2, 3, 4, codewords $c_{11}, \ldots, c_{14}$ are transmitted along a direction that is orthogonal to the directions of other codewords. Also codewords $s_{11}, \ldots, s_{14}$ are transmitted along a direction that is orthogonal to the directions of other codewords. So we can multiply $\mathbf{y}^1_1$ by $\frac{(\mathbf{H}^1_{11}(1))^\dagger}{|\mathbf{H}^1_{11}(1)|}$, multiply $(\mathbf{y}^2_1)^*$ by $\frac{(\mathbf{H}^2_{11}(1))^T}{|\mathbf{H}^2_{11}(1)|}$, multiply $\mathbf{y}^3_1$ by $\frac{(\mathbf{H}^3_{11}(1))^\dagger}{|\mathbf{H}^3_{11}(1)|}$ and multiply $\mathbf{y}^4_1$ by $\frac{(\mathbf{H}^4_{11}(1))^T}{|\mathbf{H}^4_{11}(1)|}$ to remove the interference. The result will be

$$\overline{\mathbf{y}}_1 = \overline{\mathbf{H}}_1 \begin{pmatrix} c_{11} \\ c_{12} \\ c_{13} \\ c_{14} \end{pmatrix} + \overline{\mathbf{n}}_1 \tag{5.42}$$

where

$$\overline{\mathbf{H}}_1 = \begin{pmatrix} a & \alpha^1_{11}a & \alpha^1_{12}a & \alpha^1_{13}a \\ (\alpha^2_{11})^* & -b & (\alpha^2_{13})^*b & -(\alpha^2_{12})^*b \\ \alpha^3_{12}c & \alpha^3_{13}c & c & \alpha^3_{11}c \\ (\alpha^4_{13})^*d & -(\alpha^4_{12})^*d & (\alpha^4_{11})^*d & -d \end{pmatrix} \tag{5.43}$$

$$a = ||\mathbf{H}^1_{11}(1)||_F, \, b = ||\mathbf{H}^2_{11}(1)||_F, \, c = ||\mathbf{H}^3_{11}(1)||_F, \, d = ||\mathbf{H}^4_{11}(1)||_F \tag{5.44}$$

$$\overline{\mathbf{y}}_1 = \begin{pmatrix} \frac{(\mathbf{H}^1_{11}(1))^\dagger}{|\mathbf{H}^1_{11}(1)|}\mathbf{y}^1_1 \\ \frac{\mathbf{H}^2_{11}(1)^T}{|\mathbf{H}^2_{11}(1)|}(\mathbf{y}^2_1)^* \\ \frac{(\mathbf{H}^3_{11}(1))^\dagger}{|\mathbf{H}^3_{11}(1)|}\mathbf{y}^3_1 \\ \frac{(\mathbf{H}^4_{11}(1))^T}{|\mathbf{H}^4_{11}(1)|}(\mathbf{y}^4_1)^* \end{pmatrix}, \, \overline{\mathbf{n}}_1 = \begin{pmatrix} \frac{(\mathbf{H}^1_{11}(1))^\dagger}{|\mathbf{H}^1_{11}(1)|}\mathbf{n}^1_1 \\ \frac{(\mathbf{H}^2_{11}(1))^T}{|\mathbf{H}^2_{11}(1)|}(\mathbf{n}^2_1)^* \\ \frac{(\mathbf{H}^3_{11}(1))^\dagger}{|\mathbf{H}^3_{11}(1)|}\mathbf{n}^3_1 \\ \frac{(\mathbf{H}^4_{11}(1))^T}{|\mathbf{H}^4_{11}(1)|}(\mathbf{n}^4_1)^* \end{pmatrix} \tag{5.45}$$

If α^t_{1j}, $j = 1, 2, 3$, $t = 1, 2, 3, 4$, are all real, from (5.43), it is easy to see that the equivalent channel matrix $\overline{\mathbf{H}}_1$ is real. So if QAM is used, Eq. (5.42) is equivalent to the following two equations

$$\text{Real}\{\overline{\mathbf{y}}_1\} = \overline{\mathbf{H}}_1 \begin{pmatrix} c_{11R} \\ c_{12R} \\ c_{13R} \\ c_{14R} \end{pmatrix} + \text{Real}\{\overline{\mathbf{n}}_1\} \tag{5.46}$$

$$\text{Imag}\{\overline{\mathbf{y}}_1\} = \overline{\mathbf{H}}_1 \begin{pmatrix} c_{11I} \\ c_{12I} \\ c_{13I} \\ c_{14I} \end{pmatrix} + \text{Imag}\{\overline{\mathbf{n}}_1\} \tag{5.47}$$

Then we can use the Maximum-Likelihood method to detect the real and imaginary parts of these 4 codewords separately. For example, by Eq. (5.46), we can detect $c_{11R}, \ldots, c_{14R}$ by

$$\begin{pmatrix} \widehat{c}_{11R} \\ \widehat{c}_{12R} \\ \widehat{c}_{13R} \\ \widehat{c}_{14R} \end{pmatrix} = \arg \max_{c_{11R}, c_{12R}, c_{13R}, c_{14R}} \left\| \text{Real}\{\overline{\mathbf{y}}_1\} - \overline{\mathbf{H}}_1 \begin{pmatrix} c_{11R} \\ c_{12R} \\ c_{13R} \\ c_{14R} \end{pmatrix} \right\|_F^2 \tag{5.48}$$

Similarly, using Eq. (5.47), we can detect $\begin{pmatrix} c_{11I} \\ c_{12I} \\ c_{13I} \\ c_{14I} \end{pmatrix}$. In this way, we can decode $c_{11}, \ldots, c_{14}$. Note that the decoding complexity is pair-wise. Similarly, we can remove the interference for $\mathbf{C}_2$, $\mathbf{S}_1$, $\mathbf{S}_2$ and complete the decoding. So far, we have shown how to achieve interference-free transmission and how to decode with low complexity at the receiver. In the next section, we will show how to achieve achieve full diversity for all transmitted codewords based on our interference-free transmission scheme.

5.4 Complete Precoding Scheme to Achieve Full Diversity

From the above two sections, we know how to design precoders to achieve interference-free transmission as shown in Fig. 5.3. However, the above precoding and decoding scheme can only achieve interference-free transmission, full diversity is not guaranteed.

In this section, based on the interference-free transmission scheme, we give the complete precoding design procedure for 4 time slots, which can also provide full diversity besides the interference-free transmission. First, we assume that $\mathbf{H}_1$, $\mathbf{H}_2$, $\mathbf{G}_1$, $\mathbf{G}_2$ have the following singular value decompositions

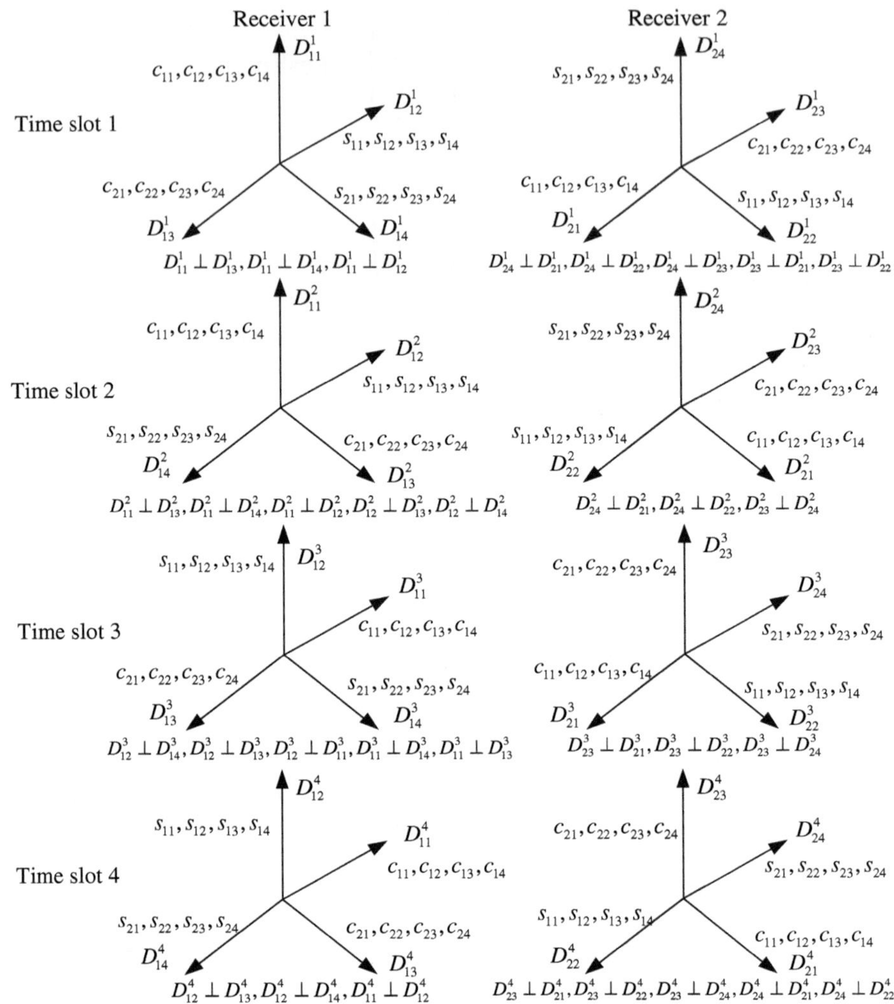

Fig. 5.4 Illustration of our precoding scheme at 4 time slots

$$\mathbf{H}_1 = \mathbf{V}_{H1}\Lambda_{H1}\mathbf{U}_{H1}^\dagger, \quad \mathbf{H}_2 = \mathbf{V}_{H2}\Lambda_{H2}\mathbf{U}_{H2}^\dagger,$$
$$\mathbf{G}_1 = \mathbf{V}_{G1}\Lambda_{G1}\mathbf{U}_{G1}^\dagger, \quad \mathbf{G}_2 = \mathbf{V}_{G2}\Lambda_{G2}\mathbf{U}_{G2}^\dagger \tag{5.49}$$

where $\mathbf{U}_{H1}(1)$, $\mathbf{U}_{H2}(1)$, $\mathbf{U}_{G1}(1)$, $\mathbf{U}_{G2}(1)$ denote the eigenvectors corresponding to the largest eigenvalues of $\mathbf{H}_1$, $\mathbf{H}_2$, $\mathbf{G}_1$, $\mathbf{G}_2$, respectively. Also we use $\mathbf{x}_1 \| \mathbf{x}_2$ to denote that vector $\mathbf{x}_1$ has the same direction as vector $\mathbf{x}_2$. As shown in Fig. 5.4, our precoder design procedure can be summarized as follows:

1. At time slot 1, design precoder $\mathbf{A}_1^1$ to make $\mathbf{D}_{11} \| \mathbf{U}_{H1}(1)$. Design precoder $\mathbf{B}_1^1$ to make $\mathbf{D}_{12} \perp \mathbf{D}_{11}$. Design precoder $\mathbf{A}_2^1$ to make $\mathbf{D}_{13} \perp \mathbf{D}_{11}, \mathbf{D}_{13} \perp \mathbf{D}_{12}, \mathbf{D}_{23} \perp \mathbf{D}_{21}$,

$\mathbf{D}_{23}\perp\mathbf{D}_{22}$. Design precoder $\mathbf{B}_2^1$ to make $\mathbf{D}_{14}\perp\mathbf{D}_{11}$, $\mathbf{D}_{14}\perp\mathbf{D}_{12}$, $\mathbf{D}_{24}\perp\mathbf{D}_{21}$, $\mathbf{D}_{24}\perp\mathbf{D}_{22}$, $\mathbf{D}_{24}\perp\mathbf{D}_{23}$.

2. At time slot 2, design precoder $\mathbf{B}_1^2$ to make $\mathbf{D}_{12}\|\mathbf{U}_{G1}(1)$. Design precoder $\mathbf{A}_1^2$ to make $\mathbf{D}_{11}\perp\mathbf{D}_{12}$. Design precoder $\mathbf{B}_2^2$ to make $\mathbf{D}_{14}\perp\mathbf{D}_{11}$, $\mathbf{D}_{14}\perp\mathbf{D}_{12}$, $\mathbf{D}_{24}\perp\mathbf{D}_{21}$, $\mathbf{D}_{24}\perp\mathbf{D}_{22}$. Design precoder $\mathbf{A}_2^2$ to make $\mathbf{D}_{13}\perp\mathbf{D}_{12}$, $\mathbf{D}_{13}\perp\mathbf{D}_{11}$, $\mathbf{D}_{23}\perp\mathbf{D}_{21}$, $\mathbf{D}_{23}\perp\mathbf{D}_{22}$, $\mathbf{D}_{23}\perp\mathbf{D}_{24}$.

3. At time slot 3, design precoder $\mathbf{A}_2^3$ to make $\mathbf{D}_{23}\|\mathbf{U}_{H2}(1)$. Design precoder $\mathbf{B}_2^3$ to make $\mathbf{D}_{24}\perp\mathbf{D}_{23}$. Design precoder $\mathbf{A}_1^3$ to make $\mathbf{D}_{11}\perp\mathbf{D}_{13}$, $\mathbf{D}_{11}\perp\mathbf{D}_{14}$, $\mathbf{D}_{21}\perp\mathbf{D}_{23}$, $\mathbf{D}_{21}\perp\mathbf{D}_{24}$. Design precoder $\mathbf{B}_1^3$ to make $\mathbf{D}_{12}\perp\mathbf{D}_{13}$, $\mathbf{D}_{12}\perp\mathbf{D}_{14}$, $\mathbf{D}_{12}\perp\mathbf{D}_{11}$, $\mathbf{D}_{22}\perp\mathbf{D}_{23}$, $\mathbf{D}_{22}\perp\mathbf{D}_{24}$.

4. At time slot 4, design precoder $\mathbf{B}_2^4$ to make $\mathbf{D}_{24}\|\mathbf{U}_{G2}(1)$. Design precoder $\mathbf{A}_2^4$ to make $\mathbf{D}_{23}\perp\mathbf{D}_{24}$. Design precoder $\mathbf{B}_1^4$ to make $\mathbf{D}_{12}\perp\mathbf{D}_{13}$, $\mathbf{D}_{12}\perp\mathbf{D}_{14}$, $\mathbf{D}_{22}\perp\mathbf{D}_{23}$, $\mathbf{D}_{22}\perp\mathbf{D}_{24}$. Design precoder $\mathbf{A}_1^4$ to make $\mathbf{D}_{11}\perp\mathbf{D}_{12}$, $\mathbf{D}_{11}\perp\mathbf{D}_{13}$, $\mathbf{D}_{11}\perp\mathbf{D}_{14}$, $\mathbf{D}_{21}\perp\mathbf{D}_{23}$, $\mathbf{D}_{21}\perp\mathbf{D}_{24}$.

Note that the design method at each time slot is similar. The key is that we change the design order for $\mathbf{C}_1$, $\mathbf{S}_1$, $\mathbf{C}_2$, $\mathbf{S}_2$ at different time slots. At time slot 1, we should design precoder for $\mathbf{C}_1$, then for $\mathbf{S}_1$, then for $\mathbf{C}_2$, finally for $\mathbf{S}_2$. At time slot 2, we should design precoder for $\mathbf{S}_1$, then for $\mathbf{C}_1$, then for $\mathbf{S}_2$, finally for $\mathbf{C}_2$. At time slot 3, we should design precoder for $\mathbf{C}_2$, then for $\mathbf{S}_2$, then for $\mathbf{C}_1$, finally for $\mathbf{S}_1$. At time slot 4, we should design precoder for $\mathbf{S}_2$, then for $\mathbf{C}_2$, then for $\mathbf{S}_1$, finally for $\mathbf{C}_1$.

In what follows, we prove that our proposed scheme can provide full diversity for each codeword. We only provide the proof for codewords c_{11}, c_{12}, c_{13}, c_{14}. The proof for other codewords is similar. The diversity is defined as

$$d = -\lim_{\rho\to\infty}\frac{\log P_e}{\log\rho} \tag{5.50}$$

where ρ denotes the SNR and P_e represents the probability of error. If we let

$$\mathbf{e} = \begin{pmatrix} e_1 \\ e_2 \\ e_3 \\ e_4 \end{pmatrix} = \begin{pmatrix} c_{11} \\ c_{12} \\ c_{13} \\ c_{14} \end{pmatrix} - \begin{pmatrix} \widehat{c}_{11} \\ \widehat{c}_{12} \\ \widehat{c}_{13} \\ \widehat{c}_{14} \end{pmatrix}$$

denote the error vector, based on Eq. (5.42), the pairwise error probability (PEP) for c_{11}, c_{12}, c_{13}, c_{14} can be written as [2]

$$P(\mathbf{c}\to\overline{\mathbf{c}}|\overline{\mathbf{H}}_1) = Q\left(\sqrt{\frac{\rho\|\overline{\mathbf{H}}_1\mathbf{e}\|_F^2}{4}}\right)$$

$$= Q\left(\sqrt{\frac{\rho\mathbf{e}^\dagger(\overline{\mathbf{H}}_1)^\dagger\overline{\mathbf{H}}_1\mathbf{e}}{4}}\right) \leq \exp\left(-\frac{\rho\mathbf{e}^\dagger(\overline{\mathbf{H}}_1)^\dagger\overline{\mathbf{H}}_1\mathbf{e}}{4}\right)$$

$$= \exp\left(-\frac{\rho\lambda}{4}\right) \tag{5.51}$$

where

$$\lambda = ||\mathbf{H}_{11}^1(1)||_F^2 |e_1 + \alpha_{11}^1 e_2 + \alpha_{12}^1 e_3 + \alpha_{13}^1 e_4|^2$$
$$+||\mathbf{H}_{11}^2(1)||_F^2 |\alpha_{11}^2 e_1 - e_2 + \alpha_{12}^2 e_3 - \alpha_{13}^2 e_4|^2$$
$$+||\mathbf{H}_{11}^3(1)||_F^2 |\alpha_{12}^3 e_1 + \alpha_{13}^3 e_2 + e_3 + \alpha_{11}^3 e_4|^2$$
$$+||\mathbf{H}_{11}^4(1)||_F^2 |\alpha_{13}^4 e_1 - \alpha_{12}^4 e_2 + \alpha_{11}^4 e_3 - e_4|^2 \tag{5.52}$$

Since

$$||\mathbf{H}_{11}^1(1)||_F^2 \geq \frac{||\mathbf{H}_1||_F^2}{4} \cdot \frac{1}{1 + |\alpha_{11}^1|^2 + |\alpha_{12}^1|^2 + |\alpha_{13}^1|^2} \tag{5.53}$$

Inequality (5.51) can be written as

$$P(\mathbf{c} \to \bar{\mathbf{c}}|\bar{\mathbf{H}}_1) \leq \exp\left(-\frac{\rho\lambda}{4}\right) \exp\left(-\frac{\rho||\mathbf{H}_1||_F^2 |e_1 + \alpha_{11}^1 e_2 + \alpha_{12}^1 e_3 + \alpha_{13}^1 e_4|^2}{16(1 + |\alpha_{11}^1|^2 + |\alpha_{12}^1|^2 + |\alpha_{13}^1|^2)}\right) \tag{5.54}$$

Therefore, we have

$$P(\mathbf{c} \to \bar{\mathbf{c}}) = E[P(\mathbf{c} \to \bar{\mathbf{c}}|\bar{\mathbf{H}}_1)]$$
$$= E\left[\exp\left(-\frac{\rho||\mathbf{H}_1||_F^2 |e_1 + \alpha_{11}^1 e_2 + \alpha_{12}^1 e_3 + \alpha_{13}^1 e_4|^2}{16(1 + |\alpha_{11}^1|^2 + |\alpha_{12}^1|^2 + |\alpha_{13}^1|^2)}\right)\right]$$
$$= \frac{1}{\prod_{j=1}^{16}[1 + \frac{\rho\tau}{16}]} \tag{5.55}$$

where

$$\tau = \frac{|e_1 + \alpha_{11}^1 e_2 + \alpha_{12}^1 e_3 + \alpha_{13}^1 e_4|^2}{1 + |\alpha_{11}^1|^2 + |\alpha_{12}^1|^2 + |\alpha_{13}^1|^2} \tag{5.56}$$

At high SNR region, (5.55) can be written as

$$P(\mathbf{c} \to \bar{\mathbf{c}}) \leq \left(\frac{\rho\tau}{16}\right)^{-16} \tag{5.57}$$

So the diversity is 16, full diversity, as long as $\tau \neq 0$. Also the coding gain is affected by τ and we can choose α_{11}^1, α_{12}^1, α_{13}^1 properly to maximize τ. The best choice for parameters α_{11}^1, α_{12}^1, α_{13}^1 depends on the adopted constellation. Such an optimization is a straightforward optimization that has been discussed in many existing literature [3]. Similarly, we can prove that the diversity for other codewords is also full.

5.5 Precoding Design for General N and M

In the last 3 sections, we have provided the precoding and decoding scheme for 2 transmitters each with 6 transmit antennas and 2 receivers each with 4 receive antennas. In this section, we will extend our scheme to a general case with any N and M.

5.5.1 $M \geq 4$

When $M = 4$ and $N = 6$, we have already provided a scheme in Sect. 5.2. From Eqs. (5.30–5.35), we know that in order to form the orthogonal structure as shown in Fig. 5.3, we need at least 6 transmit antennas. Because there are 6 equations to be solved, 6 transmit antennas will lead to 6 unknown parameters which can be solved. When $N > 6$, we will have 6 equations and more than 6 unknown parameters. We can always find the solution to form the orthogonal structure. Also with more degree of freedoms, we can achieve better coding gain.

When $M > 4$, the dimension of each signal vector at the receiver is $M > 4$ instead of 4. However, in order to achieve the orthogonal structure, we still have at most 6 equations as shown in Eqs. (5.30–5.35). So we need 6 transmit antennas for each user because each transmit antenna will lead to one unknown parameter in the precoder matrix. And the precoder design procedures are exactly the same as that of $M = 4$ in Sect. 5.2.

5.5.2 $M = 3$

A special case is when receivers have $M = 3$ antennas resulting in a 3-dimensional signal vector space, the signal vector space at the receiver. In this case, we still have 4 signal vectors. Two of these 4 vectors are useful signals and the other two are interference. So we cannot create the orthogonal structure shown in Fig. 5.3. Instead, we have to align the two interference vectors along the same direction. In this way, we will have 3 different signal directions in this 3-dimensional space. What we need to do is to make the 2 useful signal vectors orthogonal to each other and orthogonal to the interference direction as shown in Fig. 5.5. This is the main idea to achieve interference-free transmission in this case. We call this method Scheme II.

Now we show that this idea is achievable and calculate the minimum number of needed transmit antennas. Note that we design the precoders for C_1, C_2, S_1, S_2 one by one. Like before, designing precoders for the last codeword, S_2, has to satisfy the most number of constraints and results in determining the number of needed transmit antennas. If we have enough transmit antennas to successfully design the precoder for S_2, we are guaranteed to be able to design precoders for C_1, C_2, S_1 as they need to

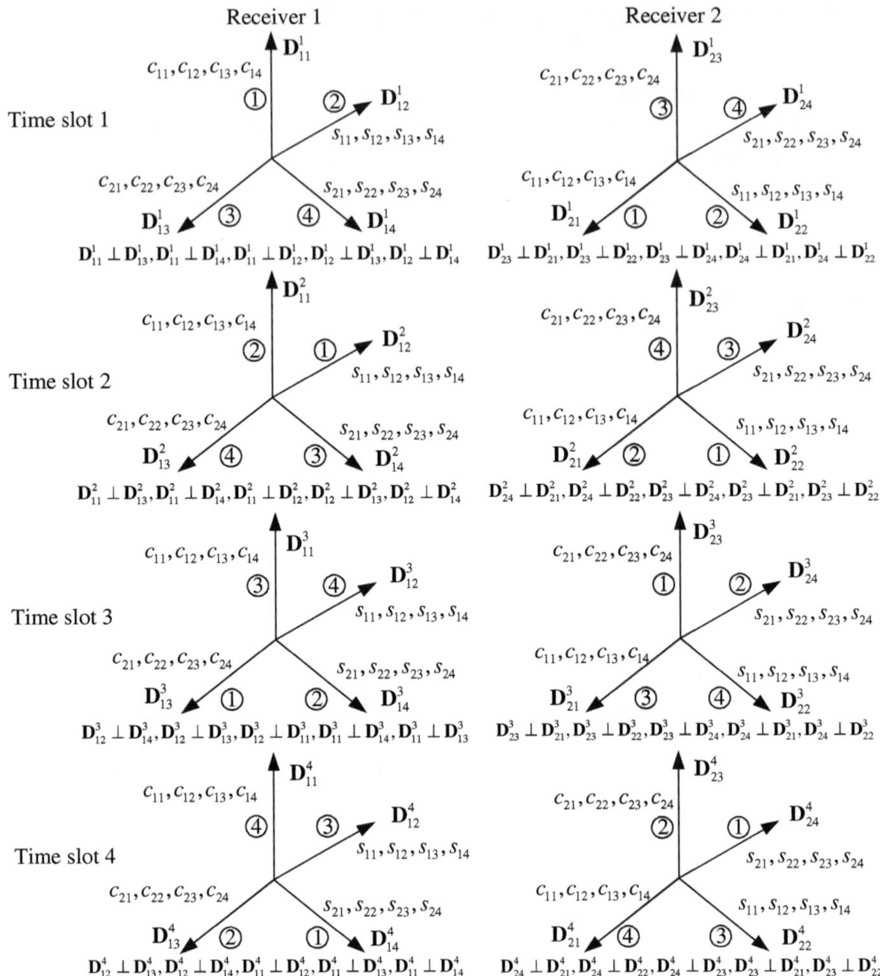

Fig. 5.5 Illustration of our precoding scheme at 4 time slots. The number in the circle denotes the design order in each time slot

satisfy less number of constraints. Therefore, in order to find the minimum number of needed transmit antennas, we only need to calculate how many transmit antennas are needed to send $\mathbf{S}_2$ without interference.

Assume that we have already finished the precoder design for $\mathbf{C}_1$, $\mathbf{C}_2$, $\mathbf{S}_1$. Then as shown in Fig. 5.5, at Receiver 1, we have $\mathbf{D}_{11}$ and $\mathbf{D}_{12}$ which are useful signals and $\mathbf{D}_{13}$ which is interference. At receiver 2, we have $\mathbf{D}_{23}$ which is useful signal and $\mathbf{D}_{21}$ and $\mathbf{D}_{22}$ which are interference and along the same direction. Now we design precoder $\mathbf{B}_2^1$ for codeword $\mathbf{S}_2$. From Fig. 5.5, we know that we need $\mathbf{D}_{14}||\mathbf{D}_{13}$, $\mathbf{D}_{24}\perp\mathbf{D}_{21}$, $\mathbf{D}_{24}\perp\mathbf{D}_{23}$, i.e.,

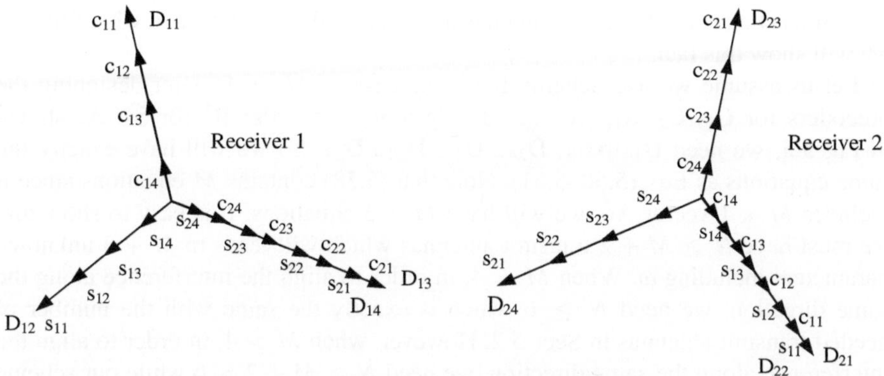

Fig. 5.6 Orthogonal structure when there are 3 receive antennas

$$\mathbf{G}_1\mathbf{B}_2^1(1) = \omega\mathbf{H}_1\mathbf{A}_2^1(1) \tag{5.58}$$

$$\left(\mathbf{G}_2\mathbf{B}_2^1(1)\right)^{\dagger}\mathbf{H}_2\mathbf{A}_1^1(1) = 0 \tag{5.59}$$

$$\left(\mathbf{G}_2\mathbf{B}_2^1(1)\right)^{\dagger}\mathbf{H}_2\mathbf{A}_2^1(1) = 0 \tag{5.60}$$

$$\|\mathbf{B}_2^1(1)\|_F^2 = \frac{1}{2(1 + (\beta_{21}^1)^2 + (\beta_{22}^1)^2 + (\beta_{23}^1)^2)} \tag{5.61}$$

Note that (5.58) contains three equations since it includes 3×1 vectors. Therefore, we need to satisfy 6 equations. One of the unknowns is the parameter ω. Thus, we need at least 5 transmit antennas since each transmit antenna will lead to one unknown parameter in the precoder matrix. To summarize, when $M = 3$, we need $N \geq 5$.

Now we provide the complete precoder design procedure for $M = 3$ and $N \geq 5$:

1. At time slot 1, design precoder $\mathbf{A}_1^1$ to make $\mathbf{D}_{11}\|\mathbf{U}_{H1}(1)$. Design precoder $\mathbf{B}_1^1$ to make $\mathbf{D}_{12}\perp\mathbf{D}_{11}$, $\mathbf{D}_{22}\|\mathbf{D}_{21}$. Design precoder $\mathbf{A}_2^1$ to make $\mathbf{D}_{13}\perp\mathbf{D}_{11}$, $\mathbf{D}_{13}\perp\mathbf{D}_{12}$, $\mathbf{D}_{23}\perp\mathbf{D}_{21}$. Design precoder $\mathbf{B}_2^1$ to make $\mathbf{D}_{14}\|\mathbf{D}_{13}$, $\mathbf{D}_{24}\perp\mathbf{D}_{21}$, $\mathbf{D}_{24}\perp\mathbf{D}_{23}$.
2. At time slot 2, design precoder $\mathbf{B}_1^2$ to make $\mathbf{D}_{12}\|\mathbf{U}_{G1}(1)$. Design precoder $\mathbf{A}_1^2$ to make $\mathbf{D}_{11}\perp\mathbf{D}_{12}$, $\mathbf{D}_{21}\|\mathbf{D}_{22}$. Design precoder $\mathbf{B}_2^2$ to make $\mathbf{D}_{14}\perp\mathbf{D}_{11}$, $\mathbf{D}_{14}\perp\mathbf{D}_{12}$, $\mathbf{D}_{24}\perp\mathbf{D}_{21}$. Design precoder $\mathbf{A}_2^2$ to make $\mathbf{D}_{13}\|\mathbf{D}_{14}$, $\mathbf{D}_{23}\perp\mathbf{D}_{21}$, $\mathbf{D}_{23}\perp\mathbf{D}_{24}$.
3. At time slot 3, design precoder $\mathbf{A}_2^3$ to make $\mathbf{D}_{23}\|\mathbf{U}_{H2}(1)$. Design precoder $\mathbf{B}_2^3$ to make $\mathbf{D}_{24}\perp\mathbf{D}_{23}$, $\mathbf{D}_{14}\|\mathbf{D}_{13}$. Design precoder $\mathbf{A}_1^3$ to make $\mathbf{D}_{11}\perp\mathbf{D}_{13}$, $\mathbf{D}_{21}\perp\mathbf{D}_{23}$, $\mathbf{D}_{21}\perp\mathbf{D}_{24}$. Design precoder $\mathbf{B}_1^3$ to make $\mathbf{D}_{12}\perp\mathbf{D}_{13}$, $\mathbf{D}_{12}\perp\mathbf{D}_{11}$, $\mathbf{D}_{22}\|\mathbf{D}_{21}$.
4. At time slot 4, design precoder $\mathbf{B}_2^4$ to make $\mathbf{D}_{24}\|\mathbf{U}_{G2}(1)$. Design precoder $\mathbf{A}_2^4$ to make $\mathbf{D}_{23}\perp\mathbf{D}_{24}$, $\mathbf{D}_{13}\perp\mathbf{D}_{14}$. Design precoder $\mathbf{B}_1^4$ to make $\mathbf{D}_{12}\perp\mathbf{D}_{13}$, $\mathbf{D}_{22}\perp\mathbf{D}_{24}$, $\mathbf{D}_{22}\perp\mathbf{D}_{23}$. Design precoder $\mathbf{A}_1^4$ to make $\mathbf{D}_{11}\perp\mathbf{D}_{12}$, $\mathbf{D}_{11}\perp\mathbf{D}_{13}$, $\mathbf{D}_{21}\|\mathbf{D}_{22}$.

Here we need to point out that when $M \geq 4$, we can also align all the interference along the same direction and use Scheme II to achieve our goal. However, we will need at least as many transmit antennas as we needed in Sect. 5.2. In what follows, we will show this fact.

Let us assume we use Scheme II for the case of $M \geq 4$. After designing the precoders for $\mathbf{C}_1$, $\mathbf{C}_2$, $\mathbf{S}_1$, we consider designing precoder $\mathbf{B}_2^1$ for $\mathbf{S}_2$. As shown in Fig. 5.5, we need $\mathbf{D}_{14}\|\mathbf{D}_{13}$, $\mathbf{D}_{24}\perp\mathbf{D}_{21}$, $\mathbf{D}_{24}\perp\mathbf{D}_{23}$. So we will have exactly the same equations as Eqs. (5.58–5.61). Note that (5.58) contains M equations since it includes $M \times 1$ vector. So, we will have $M + 3$ equations. It is easy to show that we must have $N \geq M + 2$ transmit antennas which will leads to $M + 3$ unknown parameters including ω. When $M = 4$, in order to align the interference along the same direction, we need $N \geq 6$ which is exactly the same with the number of needed transmit antennas in Sect. 5.2. However, when $M > 4$, in order to align the interference along the same direction, we need $N \geq M + 2 > 6$ while our scheme proposed in Sect. 5.2 only needs 6 transmit antennas. Therefore, when $M \geq 4$, we prefer Scheme I over Scheme II.

5.5.3 $M < 3$

When $M < 3$, the signal vector space at the receiver is 2-dimensional. But we have 4 signal vectors including 2 useful signal vectors and 2 interference signal vectors. Even if we align the 2 interference vectors along the same direction, we still have 3 signal vectors in this 2-dimensional space. Therefore, we cannot achieve interference-free transmission in this case.

In summary, when there are 2 transmitters each with N transmit antennas and 2 receivers each with M receive antennas, we can achieve interference-free transmission and full diversity simultaneously for each user if N and M satisfy the following conditions:

1. When $M = 3$, as long as $N \geq 5$, we can achieve our goal using Scheme II, i.e., by putting all interference in the same direction and making all useful signal vectors orthogonal to this interference direction.
2. When $M \geq 4$, as long as $N \geq 6$, we can achieve our goal using Scheme I, i.e., by putting all interference in a subspace which is orthogonal to the useful signal vectors as shown in Sect. 5.2.

5.6 Extension to J_t Transmitters Each with N Antennas and J_r Receivers Each with M Antennas

In this section, we will extend our previous results to a more general case, i.e., J_t transmitters each with N transmit antennas and J_r receivers each with M receive antennas. First, we provide our main result:

When there are J_t transmitters each with N transmit antennas and J_r receivers each with M receive antennas, we can achieve interference-free transmission and full diversity simultaneously for each user if N and M satisfy the following conditions:

1. When $J_t + 1 \leq M < J_t \cdot J_r$, as long as $N \geq M \cdot (J_r - 1) + J_t$, we can achieve our goal using Scheme II, i.e., by aligning all interference along the same direction which is orthogonal to the useful signal vectors.
2. When $M \geq J_t \cdot J_r$, as long as $N \geq J_t \cdot (2 \cdot J_r - 1)$, we can achieve our goal using Scheme I, i.e., by putting all interference in a subspace which is orthogonal to the useful signal vectors.
3. Otherwise, the proposed scheme cannot achieve our goal.

In what follows, we explain how we derive these conditions and give the complete design procedures to achieve interference-free transmission and full diversity for a general case.

5.6.1 $J_t + 1 \leq M < J_t \cdot J_r$

When $J_t + 1 \leq M < J_t \cdot J_r$, the only way to achieve interference-free transmission is to use Scheme II, i.e., align all the interference along the same direction. The reason is that, in the M-dimensional signal space of each receiver, there are $J_t \cdot J_r$ signal vectors including J_t useful signal vectors and $J_t \cdot (J_r - 1)$ interference signal vectors. If we do not use Scheme II, then each signal vector will occupy one dimension. But the total dimension of the receiver space M is smaller than the total number of signal vectors $J_t \cdot J_r$. So without aligning the interference, we do not have enough dimensions to achieve interference-free transmission for each useful signal vector.

On the other hand, we need $M \geq J_t + 1$. The reason is that, when $M < J_t + 1$, even if we align all the $J_t \cdot (J_r - 1)$ interference signal vectors along one direction, we still have $J_t + 1$ signal vectors including J_t useful signal vectors in each receiver space. Therefore, we do not have enough dimensions to achieve interference-free transmission for each useful signal vector if $M < J_t + 1$.

Now we analyze the requirement for N when $J_t + 1 \leq M < J_t \cdot J_r$. We assume that Transmitter k_t, $k_t = 1, \ldots, J_t$, transmits $\mathbf{C}_{k_t k_r}$, a $J_t J_r \times J_t J_r$ rate-one space time code at $J_t J_r$ time slots to Receiver k_r, $k_r = 1, \ldots, J_r$. In other words, at tth time slot, $t = 1, \ldots, J_t J_r$, Transmitter k_t sends the tth column of the space-time code $\mathbf{C}_{k_t k_r}$ to Receiver k_r. We apply the $N \times J_t J_r$ precoder matrix $\mathbf{A}_{k_t k_r}^t$ on $\mathbf{C}_{k_t k_r}$. Then at time slot t, Transmitter k_t sends

$$\mathbf{C}_{k_t}^t = \sum_{i=1}^{J_r} \mathbf{A}_{k_t i}^t \mathbf{C}_{k_t i}(t) \tag{5.62}$$

To satisfy the power constraint, we need

$$\sum_{i=1}^{J_r} ||\mathbf{A}_{k_t i}^t||_F^2 = 1 \tag{5.63}$$

Let $\mathbf{H}_{k_t k_r}$ denote the $M \times N$ channel matrix between Transmitter k_t and Receiver k_r. Then at Receiver k_r and time slot t, the received signal is

$$\mathbf{y}_{k_r}^t = \sum_{j=1}^{J_t} \sum_{i=1}^{J_r} \mathbf{H}_{jk_r} \mathbf{A}_{ji}^t \mathbf{C}_{ji}(t) \tag{5.64}$$

First, we let each symbol of $\mathbf{C}_{ji}(t)$ transmit along the same direction as we did in Sect. 5.2. So we only need to determine one column of each precoder since each column differs from other columns by certain coefficients. In other words, we have N unknown parameters. At the signal vector space of each receiver, there are $J_t J_r$ signal vectors including J_t useful signal vectors and $J_t(J_r - 1)$ interference signal vectors. From the previous discussion, we know that we only need to consider the precoder $\mathbf{A}_{J_t J_r}^t$ for $\mathbf{C}_{J_t J_r}$ assuming that we have finished the design of precoders for $\mathbf{C}_{k_t k_r}, k_t = 1, \ldots, J_t, k_r = 1, \ldots, J_r - 1$.

At Receiver k_r, $k_r = 1, \ldots, J_r - 1$, the signal vector of $\mathbf{C}_{J_t J_r}$ is interference to the receiver. Therefore, its direction should be aligned with the existing interference. Since the interference signal vector is $M \times 1$, at each of the $J_r - 1$ receivers, in order to align the signal vector of $\mathbf{C}_{J_t J_r}$ along the interference direction, we need M equations. So, we will need to satisfy $M \cdot (J_r - 1)$ equations to achieve our alignment goal in all receivers.

At Receiver J_r, the signal vector of $\mathbf{C}_{J_t J_r}$ is useful signal to the receiver. Based on our design strategy, its direction should be orthogonal to all other signal vectors. Since all interference signals are aligned along the same direction and there are already $J_t - 1$ useful signals in the space, we need to satisfy J_t equations.

Therefore, in order to solve all the above $M \cdot (J_r - 1) + J_t$ equations, we need $M \cdot (J_r - 1) + J_t$ transmit antennas which lead to $M \cdot (J_r - 1) + J_t$ unknown parameters, i.e., $N \geq M \cdot (J_r - 1) + J_t$.

5.6.2 $M \geq J_t \cdot J_r$

In this situation, we can use either Scheme II or Scheme I. We will show that Scheme II requires more transmit antennas compared to Scheme I. So we will choose Scheme I.

First, we consider Scheme I. Similar to previous cases, we only need to consider the precoder $\mathbf{A}_{J_t J_r}^t$ for $\mathbf{C}_{J_t J_r}$ assuming that we have finished the design of precoders for $\mathbf{C}_{k_t k_r}, k_t = 1, \ldots, J_t, k_r = 1, \ldots, J_r - 1$.

At Receiver k_r, $k_r = 1, \ldots, J_r - 1$, the signal vector of $\mathbf{C}_{J_t J_r}$ is interference to the receiver. Therefore, its direction should be orthogonal to the useful signal vectors.

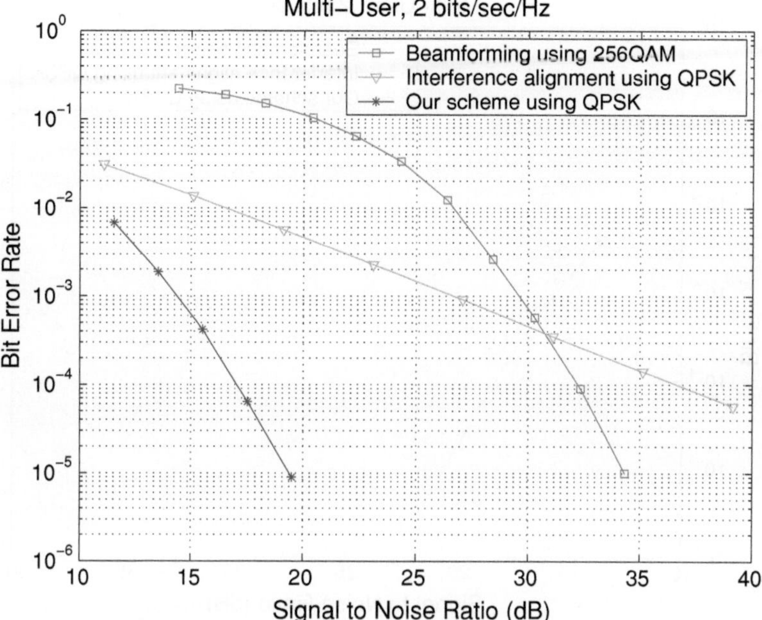

Fig. 5.7 Simulation results for 2 users each with 6 transmit antennas and 2 receivers each with 4 receive antennas

Each orthogonal relationship corresponds to one equation. Since at each of the $J_r - 1$ receivers, there are J_t useful $J_t J_r \times 1$ signal vectors, we need to satisfy $J_t \cdot (J_r - 1)$ equations.

At Receiver J_r, the signal vector of $\mathbf{C}_{J_t J_r}$ is a useful signal and its direction should be orthogonal to all other signal vectors. Since there are $J_t \cdot (J_r - 1)$ interference signal vectors and $J_t - 1$ useful signal vectors in the space, we will have $J_t \cdot (J_r - 1) + J_t - 1 = J_t \cdot J_r - 1$ equations to satisfy.

Therefore, in order to solve all these equations, it is easy to see that we only need $[J_t \cdot (J_r - 1) + J_t \cdot J_r - 1] + 1 = J_t \cdot (2 \cdot J_r - 1)$ transmit antennas which lead to $J_t \cdot (2 \cdot J_r - 1)$ unknown parameters, i.e., $N \geq J_t \cdot (2 \cdot J_r - 1)$. We need one more unknown parameter to make these orthogonal equations have a solution.

Following the logic of the last section, if Scheme II is used, we need $N \geq M \cdot (J_r - 1) + J_t$. In what follows, we show that the minimum number of needed transmit antennas for Scheme II is equal to or higher than that of Scheme I, i.e. $N \geq J_t \cdot (2 \cdot J_r - 1)$. In Scheme II, $M \geq J_t \cdot J_r$, which results in $N \geq M \cdot (J_r - 1) + J_t \geq J_t \cdot J_r \cdot (J_r - 1) + J_t = J_t \cdot (J_r^2 - J_r + 1) \geq J_t \cdot (2 \cdot J_r - 1)$. Only when $J_r = 2$ and $M = J_t \cdot J_r$, both of these two methods need the same minimum number of transmit antennas. In all other cases, Scheme I will need less minimum number of transmit antennas. From another perspective, this means that when the number of transmit and receive antennas is fixed and both Scheme I and Scheme II can work,

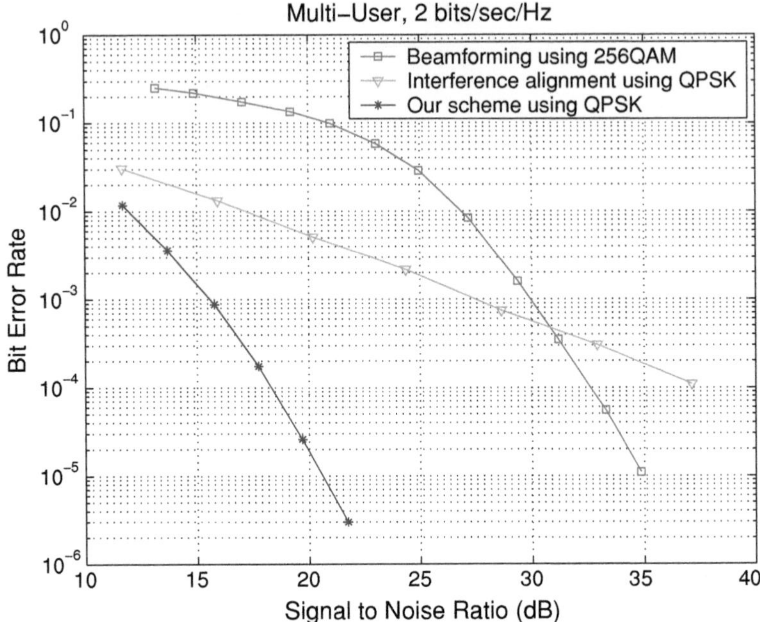

Fig. 5.8 Simulation results for 2 users each with 5 transmit antennas and 2 receivers each with 3 receive antennas

Scheme I will have more degrees of freedom to obtain better coding gain. Therefore, when $M \geq J_t \cdot J_r$, we will choose Scheme I and the required number of transmit antennas in order to achieve interference-free transmission and full diversity for each transmitter is $N \geq J_t \cdot (2 \cdot J_r - 1)$.

5.7 Simulation Results

In this section, we provide simulation results to evaluate the performance of the proposed scheme. First, we assume there are 2 transmitters each with 6 transmit antennas and 2 receivers each with 4 antennas. Then we can use Scheme I to design precoding and decoding scheme. Figure 5.7 presents simulation results using QPSK. We compare the performance of our scheme with that of two other scenarios. In the first scenario, we assume that at each time slot, only one transmitter sends signals to one receiver using beamforming. 256-QAM is used to have the same bit-rate. In the second scenario, at each time slot, each transmitter adopts the interference alignment strategy that only guarantees all the interference at any receiver are aligned along the same direction. The results show that our proposed scheme can achieve full diversity. In comparison, simple interference alignment can only achieve diversity one. Also our proposed scheme has better coding gain compared with the other two

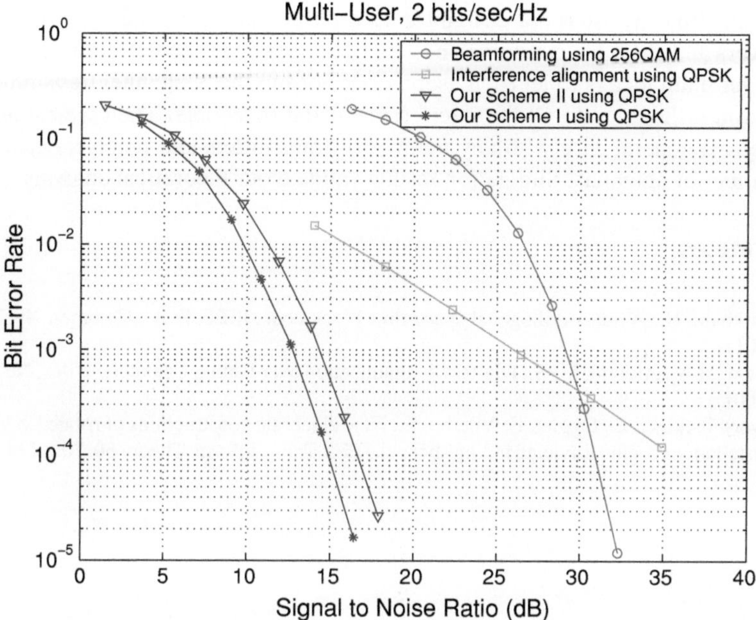

Fig. 5.9 Simulation results for 2 users each with 8 transmit antennas and 2 receivers each with 6 receive antennas

schemes. Figure 5.8 shows the results when there are 2 transmitters each with 5 transmit antennas and 2 receivers each with 3 antennas. In this case, our proposed Scheme I will not work. Instead, we can use Scheme II. Figure 5.8 shows that our proposed scheme still performs better in terms of diversity and coding gain.

Now we assume that there are 2 transmitters each with 8 transmit antennas and 2 receivers each with 6 antennas. In this case, we can use both Scheme I and Scheme II. However, as shown in Fig. 5.9, Scheme I has better performance than Scheme II. The reason is that when the number of receive antennas is the same, Scheme II requires more transmit antennas compared with Scheme I. Therefore, when the number of transmit antennas is also the same, Scheme I will have more degrees of freedom and thus have better coding gain. In addition, by Fig. 5.9, we can see that both Scheme I and Scheme II can provide interference-free transmission and full diversity.

5.8 Conclusions

In this chapter, we propose a precoding and decoding scheme for X channels to achieve interference-free transmission for each codeword with full diversity and low decoding complexity. To the best of our knowledge, this is the first scheme to achieve full diversity and interference cancellation simultaneously when all the users transmit at the same time. We start our design for a simple X channel with 2 transmitters and 2 receivers and show the conditions needed to be satisfied in order for our scheme to work. Our main idea is to let each useful codeword in the X channel transmit along a direction orthogonal to all the interference using precoders. Then we extend our scheme to a general case with any number of transmitters and receivers each with any number of antennas. Simulation results validate our theoretical analysis.

References

1. Jafarkhani, H.: A quasi-orthogonal space-time block code. IEEE Trans. Commun. **49**(1), 1–4 (2001)
2. Simon, M.K., Alouini, M.-S.: Digital Communication over Fading Channels. Wiley, New York (2000)
3. Bayer-Fluckiger, E., Oggier, F., Viterbo, E.: New algebraic constructions of rotated Z^n-lattice constellations for the Rayleigh fading channel. IEEE Trans. Inform. Theory **50**, 702–714 (2004)

MIX
Papier aus verantwortungsvollen Quellen
Paper from responsible sources
FSC® C105338

If you have any concerns about our products,
you can contact us on
ProductSafety@springernature.com

In case Publisher is established outside the EU,
the EU authorized representative is:
**Springer Nature Customer Service Center GmbH
Europaplatz 3, 69115 Heidelberg, Germany**

Printed by Libri Plureos GmbH
in Hamburg, Germany